Global Regulatory Issues for the Cosmetics Industry

Global Regulatory Issues for the Cosmetics Industry

Volume 1

Edited by

C. I. Betton
Delphic HSE Solutions Ltd, England

William Andrew

Norwich, NY, U.S.A.

Cover by Brent Beckley

ISBN: 978-0-8155-1567-8

Library of Congress Cataloging-in-Publication Data

Global regulatory issues for the cosmetics industry / edited by C.I. Betton.
     p. cm.
  Includes bibliographical references.
  ISBN 978-0-8155-1567-8 (alk. paper)
  1. Cosmetics–Law and legislation.   2. Cosmetics industry.   I. Betton, C. I.
  K3649.G58   2007
  344.04'23–dc22   2007022450

Printed and bound in the United Kingdom

This book is printed on acid-free paper.

Transferred to Digital Printing, 2010

Published by:
William Andrew Inc.
13 Eaton Avenue
Norwich, NY 13815
1-800-932-7045
www.williamandrew.com

NOTICE

Contents

Contributors   xi

Preface   xiii

Introduction   xvii

1 Risk Assessment and Cosmetics 1
C. I. Betton
1.1   Introduction: Risk Perception and Regulation   1
1.2   Risk and Cosmetics   10
1.3   The Risk Assessment Process   12
     1.3.1   Regulations   14
     1.3.2   Role of the Safety Assessor   14
     1.3.3   Guidance   15
     1.3.4   Safety Assessment   16
     1.3.5   Assessment Contents   17
     1.3.6   Other Tests   18
1.4   Conclusion   18
References   19

**2 Regulatory Developments in Canada, Japan, Australia,
China, and India 21**
Janet Winter Blaschke
2.1   Introduction   21
2.2   Canada   22
2.3   Japan   25
2.4   Australia   28
2.5   China   29
2.6   India   31
2.7   Trade Alliances   31
2.8   California   32
2.9   Conclusion   32

3 The REACH Regulation of the European Union 35
C. I. Betton
3.1   Introduction   35
3.2   Why REACH?   36
3.3   REACH: Overview   38
3.4   REACH: Aims   39

	3.4.1 Pre-registration	41
	3.4.2 Substance Information Exchange Fora (SIEF)	43
3.5	Exemptions: Cosmetics?	44
3.6	Registration	46
	References	47

4 REACH: An Example of the New Paradigm in Global Product Regulation   **49**
Felise Cooper and Kenneth Rivlin

4.1	Introduction	49
4.2	REACH: "No Data, No Market"	51
	4.2.1 What is REACH?	51
	4.2.2 How Does REACH Affect the Cosmetics Industry?	52
	4.2.3 How will REACH Work?	52
4.3	What Should Companies be Doing Now?	53
	References	54

5 Developing a Global Regulatory Strategy: Leveraging Local Knowledge to Drive Rapid Market Entry   **55**
Neil L. Wilcox

5.1	Introduction	55
5.2	Globalization: "One Size Does Not Fit All!"	56
5.3	Using Regulatory Compliance for Competitive Advantage	56
	5.3.1 Organizational Design	57
	5.3.2 Targeted Capability	58
	5.3.3 Local Knowledge	58
	5.3.4 Business Engagement	59
	5.3.5 Cross-Functional Processes, Systems, and Tools	59
	5.3.6 Communication Plan	59
5.4	The Global Launch: "Driving with Insights and Regulations"	60
5.5	Conclusions	62

6 Cosmetics: Toxicity and Regulatory Requirements in the US   **63**
Harold I. Zeliger

6.1	Introduction	63
6.2	Regulatory Requirements for Cosmetics	65
	6.2.1 Warning Statement	66
	6.2.2 Ingredient Declaration	67

6.3   California Proposition 65 and California Safe
      Cosmetics Act of 2005   67
6.4   EU Cosmetics Regulation   68
6.5   The Future of Cosmetics Regulatory Requirements   68

**7   Restricted Substances in Consumer Products:
     The Challenge of Global Chemical Compliance   71**
     Rudolf A. Overbeek and Joel Pekay
7.1   Introduction   71
7.2   Addressing the Challenges of Emerging Globally
      Restricted Substances Regulations   73
7.3   Restricted Substances: Strategy Definition   74
      7.3.1   Steps to Meeting Global Compliance
              Directives   75
7.4   Best Practices   77
7.5   Recommendation and Conclusions   79

**8   In Vitro Toxicology for Cosmetics: Regulatory
     Requirements, Biological Limitations   83**
     Ray Boughton
8.1   Introduction   83
8.2   EU Cosmetic Legislation and Animal Testing   84
8.3   A Viable Timeline?   86
8.4   Acute Toxicity   88
8.5   Skin Corrosion   92
8.6   Skin Irritation   93
8.7   Eye Irritation   95
8.8   Skin Sensitization   96
8.9   Carcinogenicity   100
8.10 Reproductive/Developmental Toxicity   101
8.11 Toxicokinetics   103
      8.11.1   Absorption   104
      8.11.2   Distribution   105
      8.11.3   Metabolism   106
      8.11.4   Excretion   108
      8.11.5   Toxicokinetics   108
8.12 Other Considerations   109
8.13 Conclusions   109
      References   111

**9   Nanotechnology and Nanomaterial Personal Care Products:
     Necessary Oversight and Recommendations   117**
     George A. Kimbrell
9.1   Introduction   117

9.2 What is Nanotechnology Anyway? A New
World of Tiny Technology 119
   9.2.1 Nano Means Fundamentally Different 121
   9.2.2 Manufactured and Engineered
Nanomaterials vs. Natural Nanoparticles 121
   9.2.3 The Next Industrial Revolution? The
Stages of Nanotechnology's Predicted
Development 122
9.3 Nanomaterials in Consumer Products: The
Future is Now 123
   9.3.1 Measures of Nanotechnology's Maturation 123
9.4 What are the Human Health Risks of Nanotechnology
and Nanomaterials in Personal Care Products? 125
   9.4.1 Nanotoxicity 126
   9.4.2 Unprecedented Mobility 127
   9.4.3 The Public at Large 128
   9.4.4 The Question of Skin Penetration 128
   9.4.5 Nanomaterial Worker and Workplace Risks 129
9.5 What are the Environmental Risks of Nanotechnology
and Nanomaterials in Personal Care Products? 130
   9.5.1 Huge EHS Unknowns 131
9.6 FDA's Regulatory Stance on Nanotechnology
and Nanomaterial Personal Care Products 132
9.7 Nanotoxicology: Nano-specific Testing Paradigms
Are Required 133
9.8 Nanomaterial Oversight Developments from FDA 135
9.9 Conclusions and Recommendations for Government
and Industry 138
   9.9.1 Support Much More Vigorous EHS Research 139
   9.9.2 Acknowledge the Unknowns and
Fundamental Differences of Nanomaterials
and Act Accordingly 139
   9.9.3 Prepare to Meet the EU Standards:
The Burden of Proof Is on Industry 140
   9.9.4 Transparency 141
   9.9.5 Repeating Past Mistakes: Running From
Your Product's Label Is Not a Business Plan 143
   9.9.6 Lifecycle 143
   9.9.7 See the Big Picture: The Question
Is Not If, But When 144
   9.9.8 Learn From The Past 144
References 145

**10   Navigating the Turbulent Waters of Global Colorant
      Regulations for Packaging   155**
      Wylie H. Royce
      10.1   Introduction   155
             10.1.1   Authors's Note   155
      10.2   Definitions   156
             10.2.1   United States   156
             10.2.2   Canada   156
             10.2.3   European Union   156
             10.2.4   Asia   156
      10.3   Regulations   157
             10.3.1   United States   157
                      10.3.1.1   Personal Care and Cosmetics   157
                      10.3.1.2   Drugs   157
                      10.3.1.3   Other Packaging Concerns   158
             10.3.2   Canada   158
                      10.3.2.1   Cosmetics and Personal Care   158
                      10.3.2.2   Drugs   159
             10.3.3   European Union   159
                      10.3.3.1   Mutual Recognition   159
                      10.3.3.2   Colorants Lists   159
                      10.3.3.3   Ensuring Safety   159
                      10.3.3.4   Other Concerns   160
      10.4   Choosing the Right Colorants for Your Product   160
             10.4.1   Dyes   160
             10.4.2   Pigments   161
             10.4.3   FD&C Colorants   161
             10.4.4   Migration   161
      10.5   Meeting the Design Challenge   161
             10.5.1   Risk Assessment   162
             10.5.2   The Quandry of Non-Regulation
                      Regulations   162
             10.5.3   Allergens   162
             10.5.4   Testing   163
      10.6   Importing Packaging   163
             10.6.1   Liability Issues   163
      10.7   Regulations at a Glance   164

Index   165

Contributors

Cliff Betton
Delphic HSE Solutions Ltd
12 Peel Avenue
Frimley, Camberley
Surrey GU16 8YT
UK

Janet Winter Blaschke
International Cosmetics and
    Regulatory Specialists, LLC
947 Manhattan Beach Boulevard
Manhattan Beach, CA 90266
USA

Ray Boughton
Intertek Testing Services
Centre Court
Meridian Business Park
Leicester LE19 1WD
UK

Felise Cooper
Global Environmental Law Group
Allen & Overy LLP
1221 Avenue of the Americas
New York, NY
USA

George A Kimbrell
International Center for Technology Assessment
Suite 302
660 Pennsylvania Avenue
Washington, DC 20003
USA

Ruud A Overbeek
Restricted Substances
Intertek
Suite 200
2200 West Loop South

Houston, TX 77027
USA

Joel Pekay
Restricted Substances
Intertek
Suite 200
2200 West Loop South
Houston, TX 77027
USA

Kenneth Rivlin
Global Environmental Law Group
Allen & Overy LLP
1221 Avenue of the Americas
New York, NY
USA

Meyer R Rosen
Interactive Consulting
PO Box 66
East Norwich
NY 11732
USA

Wylie H Royce
Royce Associates
35 Carlton Avenue East
East Rutherford, NJ 07073
USA

Neil L Wilcox
Global Regulatory and Scientific Affairs
Kimberly-Clark Corporation
2100 Winchester Road
Neenah, WI 54956
USA

Harold I Zeliger
Zeliger Chemical,
Environmental and
Toxicological Services
West Charlton, NY
USA

Preface

As Series Editor for William Andrew's "Breakthroughs in Personal Care and Cosmetic Products" Series, it is my privilege to write the Preface for this book. About 2 years ago, I was moderating a well-attended technical conference at the Health and Beauty America (HBA) Global Expo in New York City being held at the Jacob Javits Center. The speaker was just finishing his talk about some novel ingredients and there was a question from the audience about safety testing.

The speaker said something like he did not know all the details about this area but he did not think it was much of an issue. As I looked out across the room, there was a transformation that occurred. People who were tired and slouching suddenly came alert and sat up sitting on the edge of their chairs. It was as if an energy had swept into their bodies. Hands went up, one person was moving his head back and forth in the "NO" motion and suddenly a "normal" session with the expected few questions turned into about seven to ten people running up to the microphone to ask questions. I let the ensuing dialog continue for quite a while because of the level of interest; not on the speaker's topic, but on his off-handed comment about safety testing.

The discussion which ensued made me aware of several things. First, there were a few people who were talking about something called "REACH", which was legislation coming out of the EU. While a few of the attendees had heard about it, and some even seemed to know a lot about it, most of the people did not know, or worse yet, did not know that they did not know.

The dialog between the speakers and attendees escalated into a heated discussion which basically focused upon how much testing was needed for products, new ingredients, etc., and the consumer fears of "bad chemicals." It also addressed the industry's commitment to generate safe products and how to handle emerging nutraceuticals and cosmeceuticals as well as what have come to be called "functional actives" that are based upon research and a sophisticated understanding of the biochemistry of the skin.

Since that time, I have listened to many individuals asking good questions about how to test such things without animals and the impact of new regulations emerging from the 26 countries constituting the EU. These apparently required a far more stringent set of requirements for protecting the public than was set forth in US Regulations by the FDA.

Even the EU point of view of looking at the *possibility* that something might be harmful was being raised and how one could detect them at ever lower levels as newer, more sophisticated instrumentation came along. These conversations have also delved into what kind of data should be collected, what levels of materials might be dangerous and can one reliably assess the negative, long-term effects, if any. All of this was set in a context where the Federal Government's Regulations, under the FDA, had previously, years before, arbitrarily, but thoughtfully, divided the "ingredient" world into the legal "drug" and "cosmetic" partitions. Each of these has their own regulated protocols for safety testing and claim management. As we well know, even the more stringent requirements for drugs do not preclude the possibility of a material causing harm.

It may well be that these legal distinctions have become anachronistic with the passage of time and the appearance of "functional actives" which may best fit between the legal "drug" and "cosmetic" divisions. As I write this, I want you to know I am aware that I may be saying some things that are not usually the subject of wide discussion, although they are certainly not hidden, by any means. In my view, they need to be said because, it turns out that legislation from abroad has taken the first steps, to address this issue and, as one of my favorite clients would say, "there will be pushback" from our own government as well as others around the world.

Much attention has been paid to this legal distinction of "drugs" vs. "cosmetics." The regulatory requirements for testing and safety are quite different, as are the concomitant costs of obtaining such information. While the drug industry generally spends far more money to obtain such mandated information, the personal care and cosmetic industry generally spends much less, as it conforms to current legislation. Claims written on packages of products are guided by these regulations and legal specialists are often consulted as to what one can claim so as to tell something good about what a product does as opposed to not going "too far" and risking crossing the line into "drug" territory.

As a Fellow of the American College of Forensic Examiners, involved as an expert in litigation relative to personal injury and products liability, and a Fellow of the Royal Society of Chemistry (London), these certifications, responsibilities and interests, on both sides of the Atlantic sparked something in me and called me to action. Here was an area of great interest to ingredient manufacturers, suppliers of finished goods, chemical manufacturers as well as litigators and government regulators.

Here was an emerging critical area for the personal care and cosmetic industry with global, or perhaps better said, international impact. Here was an area that people needed to know about and the time for this was NOW. The Vice President of CMPi, the Global Corporation that runs the HBAExpo Educational Conference as but one of its over 20 different areas was sitting next to me at the time of this conference. He turned to me just as I turned towards him and I can still remember the look in his eyes (and perhaps he in mine). He said, "This area has Sizzle—the room is completely energized . . . this is a topic that we must step into and lead the way to educating people about what they need to know and who they can go to for support, if they need it". "My thinking exactly", I said.

And so, it began. HBA Global Expo management offered me the opportunity to organize the First HBA Regulatory Symposium 2006. Within three months I had seventeen speakers and panelists and, with almost no early advertising, for a topic that there had not been a conference on in the past. We set up a technical symposium which drew international attention from the press and industry. I had to turn away speakers and the one day session was so long that some people were willing to miss their planes and stayed over just to be at the entire conference and the final one hour professionally moderated panel discussion (by Rebecca James Gadberry, President of YG Labs and myself). We recorded the entire conference and made it available to the public by placing it on the internet in a webinar (www.hbaexpo.com).

It was not long before other "REACH" conferences came into being on both sides of the Atlantic. A Society of Cosmetic Chemists meeting in New York later that year had a technical conference in which over 230 people attended—almost twice what we had in our First HBA Regulatory Conference—and other conferences sprang up at In-Cosmetics and elsewhere. Unlike what you might think at first, the issues at hand are not dry and boring, especially for technical people. They are not just for attorneys. These issues impact everyone in the personal care and cosmetic industry from the technologists to the scientists; from the marketers to the business management; from the testing laboratories to the in-house counsel and to the outside experts, consultants, regulators, and politicians. The global press recognizes the breadth of the impact of these changes. It affects every aspect of the business.

As Series Editor of the William Andrew Series I mentioned previously, I invited Cliff Betton of Intertek, a major contributor to the conference, to edit a book on this subject based upon the conference proceedings. Happily, he accepted! He asked all of the speakers to contribute and, as things go (as I had learned from editing my 1,100 page book *Delivery System Handbook: Technology, Applications and Formulations*), many said yes and some said yes initially but, for various reasons we call "life" could not do it later. To his great credit, Cliff then found experts who had not spoken at the HBA Regulatory Summit 2006 and they filled in chapters that needed to be done.

By joining forces, William Andrew's commitment to push the book forward on the fast track and HBA's agreement to be a media partner company has resulted in this book. It is the first of an annual book mini-series surveying issues in this critical and rapidly changing area. These changes affect the health, safety and well-being of literally billions of consumers, their governments and the corporations involved in the prodigious task of not only creating novel, effective and safe products, but also complying with regulations, that vary from country to country.

The accelerating regulatory push is a result of increasing consumer pressures, public awareness and what I have now begun to think of as the "legislative ripple effect" and its flow outward from places like California, USA, and the EU, to other areas around the globe. This process has many facets, and is, I have come to see, more than a "regulatory" issue. *It is one of a fundamental shift in the definition of chemical product safety* and its impact on consumers and regulators alike.

In the intervening months, HBA Global Expo extended an invitation to me to be the Chief Scientific Advisor of their entire technical program. As part of my responsibility we have initiated the *Second International HBA Regulatory Symposium*. It will be held on September 19, 2007, in New York City at the Javits Center (www.hbaexpo.com).

I invite you to attend for there is much to learn for all of us, including how to get along better, even if we live in different countries and have different points of view.

It is my intention and promise that the content of the *Second HBA International Regulatory Symposium* will go to the heart of these current and emerging issues, from a rational and educational point of view. It will focus broadly upon the multi-aspect, dynamic interplay of technology, safety, consumer issues, education of legislators and provide support for the generation of regulations, both existing, planned, and being contemplated. It will also address the proactive initiatives being taken by the personal care and cosmetic industry to educate consumers, technologists, scientists and legislators alike in current safety information and trends.

Next year, look for Volume 2 of this mini-series. The track is laid and we are moving forward.

Meyer R. Rosen, FAIC, CPC, CChE, DABFET, CChem
Fellow of the Royal Society of Chemistry (London)
Fellow of the American College of Forensic Examiners
Chief Scientific Advisor to HBA Global Expo
President, Interactive Consulting, Inc.
(www.chemicalconsult.com)

Introduction

Each year Health and Beauty America (HBA) holds an exhibition. At this exhibition companies from all over the USA and indeed the world come to demonstrate their products, technology and to open up new avenues for business. Similar exhibitions are held annually in China, Hong Kong, Tokyo, England, France, Italy, and Russia; the list is extensive and growing annually. At each venue hundreds and at the large events such as the HBA Exhibition in New York, or the exhibition in Bologna, Italy, possibly thousands of businesses spend millions of dollars, pounds, euros, or yuan just to keep up with, and show that they are keeping up with, modern trends and technological developments. Cosmetics is a business that affects almost every family on Earth as a user.

We all keep clean, and we all, from supermodels on the catwalk, and Hollywood stars with their perfect features emblazoned on a screen 20 feet high, through "ordinary" citizens to lost tribes in the Amazonian rainforest who daub themselves with mud or ochre, try and enhance our appearance. The making of cosmetics may not be the oldest profession, but it comes close. Ancient Britons were painting themselves with woad to terrify their opponents and saw off Julius Caesar in 55 and 54 BC and it was terrifying enough to keep the Romans away for nearly a hundred years. Cosmetics are effective in what they do but up until recent times, there was no real regulation of cosmetic products, indeed there was not even a definition.

There were accidents and incidents of ill health bought on by the use of cosmetics throughout history, but given the poor lifespan of humans until comparatively recent times, the lack of safe drinking water in urban conurbations that have grown exponentially following the industrial revolution, and all the other causes of ill health and premature death, safety of cosmetics was low on the agenda.

The use of atropine from the berries of the deadly nightshade to dilate the pupils and make the user more beautiful (Bella Donna means beautiful lady in Italian and the deadly nightshade's botanical name is *Atropa belladonna*) was a long established practice. Cleopatra is reported to have used Kohl (a natural mineral consisting primarily of galena—lead sulfide) and its use is still common in many Middle Eastern countries to this day. It is considered to be harmless; nevertheless several studies have linked exposure to Kohl to the death of infants due to lead poisoning. White lead compounds are reputed to have been the basis of the white make-up used by Queen Elizabeth I in England. Perfumes also histori-cally contained a number of neurotoxic and allergenic materials such as musk ambrette.

As public health improved after the World War II and lifespans started to improve significantly, regulation started to develop. Like all good ideas it seems to have occurred in several places at once and human nature being what it is everyone had similar ideas, but not identical ones. The defini-tions of what constitutes a cosmetic are different in different places around the world. In Europe, a cosmetic is very broadly defined, the definition in the Cosmetic Products Regulation of the EU (76/768/EEC) is:

> "Cosmetic product" means any substance or preparation intended to be placed in contact with any part of the external surfaces of the human body (that is to say, the epidermis, hair system, nails, lips and external genital organs) or with the teeth and the mucous membranes of the oral cavity with a view exclusively or mainly to cleaning them, perfuming them, changing their appearance, protecting them, keeping them in good condition or correcting body odours except where such cleaning, perfum-ing, protecting, changing, keeping or correcting is wholly for the purpose of treating or preventing disease.

However, in the US an anti-dandruff shampoo is considered to be an over the counter pharmaceutical whereas in Europe it is a cosmetic and in Aus-tralia, a sunscreen lotion is classed as a "dermaceutical" whilst in Europe it is again considered to be a cosmetic. Few people would class products like an industrial hand cleaning gel, found in workshops everywhere, to be a cosmetic, but it fits the European definition perfectly.

As with definitions, different countries have different ideas on what ingre-dients should be permitted in cosmetics: products that may be used on a daily basis, on all areas of the body and remaining in contact and use for prolonged periods of time. Colours are restricted, but not all the same colours and not all in the same ways; some colours are prohibited in the USA, but allowed in Europe and vice versa; a formulator has to be aware of the markets that their products are going to be sold in if they are to be universally acceptable.

Given the expansion of regulation, the globalization of the cosmet-ics business and the need for exporters and importers to understand that simply because something is OK in one country, it is not necessarily acceptable in another, in September 2006, HBA organized the first annual Cosmetic Regulatory Forum.

This event was held at the Jacob Javitz Exhibition Centre in New York and ran in conjunction with the HBA Exhibition. Speakers were invited, mainly from the USA but also from Europe and addressed topics such as toxicology, packaging, colours, animal and non-animal testing, regulation, risk assessment, corporate governance and nanotechnology.

This book is based on the presentations that were given at the conference and will be the first in what is expected to be a continuing series of events and publications. Regulation is changing rapidly and this is reflected in the chapters of this book, indeed some regulations are changing so rapidly that one of them, REACH, did not exist when the conference took place, and was passed into law in December 2006 and comes into force on June 1, 2007. The REACH legislation will have far reaching impacts on sales and business in Europe as it is a revolutionary change in the way in which regulation is conducted (see Chapters 3, 4). There was much lobbying against REACH from countries outside of the EU; the US vice president went on a personal tour of EU capitals in an attempt to persuade governments to drop the whole idea. But REACH is with us and it has already been translated into Chinese and the Chinese government is already working on its own version of REACH. The market in China is not just huge—more than the EU and US combined—but also growing at a phenomenal rate. Legislation in one part of the world is increasingly ending up being adopted by other countries.

The aim of HBA is to have a continuing dialog and update of issues of relevance to the cosmetics industry at the annual HBA Regulatory Forum and for that dialog to be recorded in the form of a publication which will develop over time into an authoritative record and reference of all regulatory issues of relevance to the cosmetics industry world wide. This first volume addresses some fundamental issues, some of them controversial, and reflects the diverse and changing regulatory environment that cosmetics companies have to deal with. One of the most controversial areas at the moment is the use of animal testing in assessing safety. This has been prohibited by Law in the United Kingdom since 1995, but has been standard practice in other EU countries and the USA until recently when the Seventh Amendment to the EU cosmetics regulations prohibited the use of animal tests on cosmetic products and their ingredients (if solely for the purposes of demonstrating compliance to the Cosmetics Directive).

In the USA, a very different regulatory regime operates. Cosmetics must be "demonstrably safe" and this is generally proved by virtue of animal tests, although there are no specific requirements for this. It is also, however, possible to market products that have not be assessed for safety providing that they are correctly marked and that the consumer is then able to choose whether or not to buy the product.

This slim volume sets the scene for current regulation. Subsequent conferences and editions of this book will address different issues and will document the changes in legislation that will occur as more countries introduce legislation and those such as the EU and the USA adopt and adapt to each others laws in their attempts to keep up with the global market economy.

It has been an interesting process editing this volume and I would like to thank all of the authors who have contributed their knowledge and expertise to get the series off to a sound start, to Martin Scrivener at the publishers for his support and encouragement, and to Meyer Rosen for giving me the opportunity in the first place. I hope that you find this book of value and that it will be not just a volume on your shelf but something that is a reference work that helps to understand, if not the detail of the legislative environment in 2007, for that you need to go to the source documents, but gives an understanding of the basic principles which the legislations are trying to achieve and the trends that can be expected in the future. Those who can anticipate change earliest and most effectively will prosper, whilst those who think that things will stay the same will have a difficult future.

C.I. Betton
Camberley, England
July 25, 2007

1

Risk Assessment and Cosmetics

C. I. Betton

Delphic HSE Solutions Ltd
Camberley, UK

1.1 Introduction: Risk Perception and Regulation

At the Health and Beauty America (HBA) Conference in New York in 2006, the plenary paper was presented by Professor Bruce Ames. Professor Ames[1] talked about "Putting Cancer Risks into Perspective" and gave a wonderful exposition on the real risks confronting society that lead to damaged health in adults and children and compared those with the regulatory system that controls our exposure to chemicals that are regulated as a means of controlling and in some senses, eliminating risks.

The principle causes of cancer identified by Professor Ames were:

- Smoking (~30%)
- Poor diet (~35%): excess calories and obesity, lack of fiber, lack of essential micronutrients
- Chronic infectious disease (~20%), mainly in poor countries
- Hormonally mediated (~20%): breast cancer, endometrial cancers, etc.

C. I. Betton (ed.), Global Regulatory Issues for the Cosmetics Industry Vol. 1, 1–20
© 2007 William Andrew Inc.

- Occupationally derived (~2%)
- Pollution (<1%), mostly due to air pollution

Regulation of cancer is a major effort by national and international bodies in western democracies. Restrictions that control or ban the use are placed on chemicals that are associated with cancer. The European Union (EU) and the US FDA and EPA control and restrict chemical use where there are suspicions of carcinogenic potential, and yet the figures quoted by Professor Ames indicate that the main causes of disease are not related to "chemicals" at all but rather are associated with lifestyle choices; poor diet and smoking accounting for approximately 65% of all cancers. It could be argued that the almost insignificant contribution of chemical causes to the total (occupational cancers and those related to pollution) is a positive demonstration of the effectiveness of regulation, and that if regulation were to be relaxed, then the numbers would increase. That is a point of view.

Risk is always present in any human activity and acceptance of risk is a political decision, whether it is made by the individual who chooses to smoke, or to go skiing or swimming with sharks, or by politicians on behalf of society as a whole.

Assessment of risk is an emotional and complex issue. As individuals we make choices not necessarily based on facts. Individuals choose to smoke even though there are clear warnings on the packs, of death and cancer, partly because of the pleasure they derive from the drug and partly due to the long-term nature of the consequences: as my own mother used to say "Yes I know all that, but I could just as easily be knocked down by a bus tomorrow." In the UK, the National Statistics Office publishes mortality data annually, as required by law. The latest figures, published in 2006 covering deaths during 2005, show some interesting figures.[2]

During 2005 exactly 512,692 deaths were recorded. Accidents were responsible for 10,962 (~2%), of which 2,740 (0.5%) were related to transport and falls accounted for 3,006 (0.6%) of the total. Deaths due to cancers of all types were 138,454 (27%), a major cause of mortality in England and Wales. However, if we look at the age distribution of those deaths, an interesting pattern emerges, as shown in Figure 1.1.

This graph shows that the number of deaths due to cancer (all types) in England and Wales is not significant in terms of total numbers during the early years of life, but that the cause of death becomes more significant

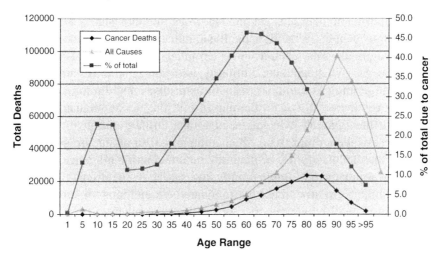

Figure 1.1 Deaths due to cancer in England and Wales during 2005 by age group of 5 years.

as we get older. However, if the numbers due to cancer are considered in terms of all causes of death, there are two periods of maximum concern. The first is among the very young, peaking at around 10 years old, where cancer represents about 22% of all causes of death. The rate then falls to a low from the mid-teens to the late twenties (the reproductive years) before rising to peak at around 60 years, where cancer accounts for over 45% of all deaths reported in England and Wales, before falling again almost linearly as we get older. We all die of something, but it seems that the odds of it being cancer diminish as we get past 60 years.

As an individual we assess risk, determining what is acceptable and what is not on a case by case basis. We are concerned about threats to our lives and health. Everyone is concerned about avoiding accidents and there are enormous resources directed at road transport: traffic police, speed cameras, and a whole infrastructure aimed at controlling accidents on the road. Yet there are still accidents and in reality the number of people killed on the road is less than those who die as a result of falling, mainly in the home. There is little regulation about how fast we may run down the stairs to answer the front door, which is not perceived as a risk, or if it is, then it is a risk we ourselves can control.

Fear of flying is a significant factor in people's perception of risk, yet in 2005 there were no deaths in England and Wales related to flight, but over

2,500 occurred on the roads, yet the fear of getting into a car is not a problem for most people. Why? It is the illusion that in a car you are in control and managing the risk, whereas in an aeroplane you are helpless and have no control over what happens. Similarly, with cancer, smokers are in control but do nothing to diminish risks to themselves, and do not care what happens to others exposed to the majority of the smoke emanating from the tip of the cigarette, whereas occupational disease and environmental exposures, while a minor issue when compared to smoking, are perceived as being beyond the ability of the individual to control and hence are perceived as a greater risk. Interestingly, awareness of the effects of smoke on nonsmokers are now much more understood and bans on smoking in public places will reduce this risk to nonsmokers, much to the chagrin of the "freedom to smoke" lobby. Cancer is a risk, it is a risk that increases as we get older, and yet the figures show that once we are past 60 years it is other things that tend to kill us, although of course we all die of something in the end. Life is a risky business.

Risk assessment as far as regulation is concerned is a political decision, and it is here that the problems start. Risks associated with chemicals are assessed based on a range of studies and the assessment of what is acceptable is based on what is "safe." There are two schools of thought on the determination of "safe": one is represented by the idea that one molecule is enough to cause cancer and that there is no such thing as a no-effect concentration; the opposing view is that for all substances and individuals there is a dose–response relationship that will result in a no-effect concentration and that changes in a body's chemistry that result from exposure are evidence that the organism is coping with the substance. Of course, it is impossible to prove a negative and thus it is not possible to prove categorically that anything is "safe," only that studies have demonstrated a "no observable effect" concentration, which is not to say that better observation would not have seen anything change nor that further studies using better techniques will not demonstrate an effect of some sort.

Faced with the uncertainties of the scientific method, regulators are between a rock and a hard place. As a society we ask them to regulate on our behalf and pressure is exerted from all corners—individuals, pressure groups, and industry—often with conflicting demands. We must also remember that politicians are human beings (it is true!). Risk is often expressed as the number of cases of disease or in the case of chemical carcinogenicity, cancer per 100,000 of the population per year. If the rate is expressed in this way, a rate of 0.1 cases per 100,000 seems low and entirely acceptable in the cold light of day as a figure; but in the UK, with a population of 60,000,000 people this would mean that six people are going to suffer adversely. It may be the case that this is fewer than those who get struck by lightening each year, far fewer than those who succumb to bee stings, nowhere near as many as those who suffer from anaphylaxis due to dietary factors such as peanuts. For those six individuals the results are catastrophic, and no politician is going to confront the grieving relatives of an affected person and admit that it is hard luck, but considered acceptable for the nation. No human being of conscience would be prepared to make such a statement.

This brings us back to Professor Ames' lecture and the regulation of risks. The perception of the population as a whole is that "chemicals" are bad and that nature is good. There is also a strong perception that we are exposed to a number of carcinogenic chemicals every day and that the health of the nation (every nation) is being put at risk for profit by the chemical industry. All of these perceptions are of course wrong, but in the world of politics and regulation, often, perception is reality.

Let us consider some facts. In the 1900s, the life expectancy for both men and women has increased from the mid-forties in 1900 to 79 years for women and 71 years for men by 1990 (US National Institute on Aging). Hardly evidence of a sick population, in an era when the output of "chemicals" has increased significantly. Professor Ames then looked at the incidence of positive findings in cancer studies, where the chemical had been tested in both rats and mice.[1] For all chemicals tested, 59% of them (377 out of 636 studied) were positive in both species, but on closer examination of the data it can be seen that both "natural" as well as synthetic chemicals have been studied. The rate for naturally occurring chemicals was 57% (86/152) and the rate for synthetics was 60% (291/484)—no real difference. A larger number of materials have been studied in just one species, with the consequence that the overall incidence of positive results is slightly lower at 52% (748/1,430), but the results in this section of results is more alarming (or reassuring depending on your point of view). Natural pesticides produced by plants, as protection against grazing by insects and other animals, show a cancer rate in rats or mice of 53% (39/73), whereas commercial synthetic pesticides (chemicals) show a rate of just 40% (79/196). An alarming finding for roasted coffee is that, 70% (21/30) of the compounds found in roasted coffee that have been studied in rats or mice are carcinogenic when tested.

Does this mean that coffee is carcinogenic? Yes and no. If given to rodents at high doses for a long time, then the majority of the components of coffee can induce tumors. Does this represent a risk to the health of humans? Again it depends on your viewpoint. Some would say that it is clear evidence of risk and that coffee should not be drunk as something has to be causing all of the cancers. But consider the exposure of the average American, French, or Italian to coffee. Given the amounts consumed and the strength of French or Italian coffee, if there was a real risk to health, then the significant increase in lifespan seen in the twentieth century could not have occurred.

From a regulatory standpoint, if regulators have not banned cigarette smoking and saved the lives of millions by doing so because it is not politically feasible to do so, how likely are they to ban the consumption of coffee where there are no measurable risks to *human* health?

The presence of carcinogens in coffee is not unique. Many foods contain materials as a part of their natural make-up that, when tested individually, contain materials that are carcinogenic in animal tests. Table 1.1 lists some of these compounds, the foods in which they are present, and their concentration ranges.

Consider also the humble tomato, constituent of most dishes in Italian cookery and a staple of the diet. Three hundred and ninety volatile chemicals have been identified as being present in tomatoes, of those 25 have been tested for carcinogenicity in experimental animals and 17 have been shown to be carcinogenic: acetaldehyde, benzaldehyde, benzene, benzyl acetate, chloroform, 1,4-dioxane, ethanol, ethylbenzene, formaldehyde, furfuryl alcohol, limonene, naphthalene, pyridine, styrene, toluene, 1,2,4-trimethyl benzene, xylene.

Does this mean that tomatoes are carcinogenic? Of course not. What it does mean is that our perception of risk from chemicals is not in line with the facts *as far as risk to humans is concerned.*

In the EU there are standardized schemes for the classification of chemicals related to their potential *hazard*. These are known as *Risk* (R) Phrases.[3] In relation to carcinogenic potential, they follow the scheme of the International Agency for Research on Cancer (IARC) based in Lyon,[4] whereby carcinogens are classified as follows:

Category 1:   Known human carcinogens
Category 2a: Proven animal carcinogens, probable human carcinogens

Table 1.1 Naturally Occurring Food Constituents That Have Been Shown to Cause Cancer in Experimental Animals[1]

Animal Carcinogen	Foodstuffs	Concentration (ppm or mg/kg)
5- and 8-Methoxypsoralen	Parsley	14
	Parsnip (cooked)	32
	Celery	0.8–25
p-Hydrazinobenzoate	Mushroom	11
Glutamyl-p-hydrazinobenzoate	Mushroom	42
Allylisothionate	Cabbage	350–590
	Radish	11
	Cauliflower	12–66
	Brussels sprout	110–1,560
	Mustard (brown)	16,000–72,000
	Horseradish	4,500
Limonene	Orange juice	31
	Mango	40
	Black pepper	8,000
Estragole	Basil	3,800
	Fennel	3,000
Safrole	Nutmeg	3,000
	Mace	10,000
	Black pepper	100
Sesamol	Sesame seed oil	75
Benzyl acetate	Basil	82
	Jasmine tea	230
	Honey	15

Category 2b: Proven animal carcinogens, possible human carcinogens
Category 3:   – Not classifiable as to human carcinogenicity

In November 2006, the UK Government published a list of materials considered to be dangerous for consumers.[5] This was based on classification of substances as carcinogens, mutagens, and teratogens using the EU scheme and based on the IARC categories listed above. What is so useful about this regulation is that it contains a table listing all Class 1 and Class 2 carcinogens, and what is even more interesting from a risk assessment perspective is the number of known human (Category 1) carcinogens there are (if you exclude petroleum products that are classified based on their composition, but *include* the chemicals that result in that classification, i.e., benzene,

1,3-butadiene and three to seven ring polycyclic aromatic hydrocarbons). The following are classified as Category 1, known human carcinogens:

- Chrome (VI) trioxide
- Zinc chromates (including zinc potassium chromate)
- Nickel monoxide, dioxide, sulfide, and subsulfide
- Dinickel trioxide
- Arsenic acid and its salts
- Arsenic oxide, dioxide, pentoxide, and diarsenic trioxide
- Triethyl arsenate
- Lead hydrogen arsenate
- 1,3-Butadiene
- Benzene
- Benzo(a)pyrene and related three to seven ring polycyclic aromatic hydrocarbons
- Vinyl chloride monomer
- Bis(chloromethyl) ether
- Chloromethyl ether
- 2-naphthylamine (β-naphthylamine) and its salts
- Benzidine and its salts
- Biphenyl-4-ylamine and its salts
- Coal tar
- Erionite
- Asbestos

A comparatively short list of chemicals *known* to cause cancer in humans, especially when compared with the number of natural and synthetic substances that have been shown to cause cancer when tested in laboratory animals.

At this point it is worth considering whether our current strategy of testing on animals and extrapolating the results directly to humans is valid, of scientific value, or whether it needs some revision.

The facts in relation to cancer appear to suggest that a revision is needed. Increasingly sophisticated measurement techniques can identify changes in biochemical processes and animal studies identify increasing number of chemicals as having carcinogenic properties, yet we are living longer and more and more we find that foodstuffs consumed by millions every day contain substances that are classified as carcinogens as a result of testing.

Then again, if we look at the disparity between the large number of chemicals identified as carcinogens in experimental animals and the comparatively small group of compounds known to cause cancer in man, it appears that there is a disconnection between the identification of hazards and the assessment of real risks to health.

Up until recent times, regulation has been based on the identification of hazards and then extrapolating those results to mankind by the application of the precautionary principle involving the use of safety factors of 10- or 100-fold. For numerous reasons, some of which have been discussed above, this system has been found wanting, in terms of relating the findings in laboratory testing to actual human experience. Animal testing has been questioned vociferously by the animal rights movements particularly in terms of its relevance to man; in some regards they have a point. Modern legislation, particularly in the EU, has, however, been moving away from identification of hazard towards a risk-based regulatory system. That is not to say, as the animal rights campaigners do so effectively, that animal tests are of no value or relevance to human health and safety; the results of all tests are important in building a profile of the properties of substances. It is the interpretation of the results that has been at fault in the past and it is important to realize that animal data, in much the same way as do data from *in vitro* tests, need careful interpretation, assessment, and extrapolation.

Animal data on chemicals is essential if they are to be handled and used safely. Animal data cannot, however, tell us what the effects of exposure to humans will be. They require careful interpretation and assessment (opinion) if effective human health protection is to be maintained and improved.

Risk-based regulation is based on opinion and this is where things become conceptually difficult for regulators, as each compound and each use of that compound has to be assessed for safety on a case by case basis. There are no general rules that can be applied, there is only the opinion of the assessor, given based on knowledge and experience of whether the risk is acceptable or not, for any given chemical being used in any given way. This is not a concept that lawyers in particular feel comfortable with; it does, however, seem to have found increasing favor with regulators in Europe. One reason perhaps being that for the first time they can remove themselves directly from the decision-making process, and pass responsibility where it rightly belongs, with manufacturers and "experts" in the complex process of analysis of biological information. This process, which began with the EU Cosmetics Directive in the 1970s will reach its culmination with the REACH (Registration, Evaluation, Authorisation and Restriction of Chemicals) regulation (see Chapter 3) when it comes into force in June 2007 and begins to have effect in 2010 and thereafter. Under REACH it is the responsibility of the manufacturer to assess risk, the regulator merely has to agree, or not, with that assessment!

1.2 Risk and Cosmetics

A purchaser of a cosmetic product has an expectation and a right to believe that the product that they are being sold will not cause them harm. It was not always the case. Atropine from belladonna (literally "beautiful lady") was used to dilate the pupils of ladies in order to make them appear more attractive to men, lead-based pigments have been used leading to lead poisoning, and a number of allergenic materials have been incorporated into perfumes. All these have had adverse effects on people's health.

In 1978, the EU began the regulation of cosmetics with the introduction of the first Cosmetics Directive.[6] This was an instruction to each member state to introduce legislation that was aimed at controlling the hazards to health by cosmetic use.

The traditional method of assessing the safety of cosmetic products, in the period just after World War II, involved the use of animal models to generate data on irritancy and general toxicity. Tests were developed for the assessment of cosmetics that became standard methods for general toxicological assessment; the Draize tests for skin and eye irritation still form a part of toxicological assessment procedures. These tests were, however, developed to discriminate between and identify potential problems with cosmetic products; if the products are to be successful in the marketplace, they should be generally of low or no irritant potential. The tests as developed were therefore necessarily stringent, and while this may not have been an ethical issue when used for the purpose that they were originally intended at that particular time, applying these same tests with little thought to a much wider range of products began to raise serious doubts about the validity of the results and the ethics of carrying out the test in the first place.

In the early 1980s, work was taking place across Europe under the aegis of the EU to give a more appropriate foundation for animal testing protocols that could be applied to a broad range of industrial chemicals, which culminated in the Dangerous Substances Directive and the European scheme for the Classification of Dangerous Substances.[3] These studies became mandatory for new chemicals after 1981. This requirement coincided with a rapid growth in political pressure caused by the activities of various animal rights groups, who rather cleverly focused on the testing of cosmetics—products that are not vital for survival and where use and exposure is a matter of choice and not necessity. The charge that the tests were carried out for profit went largely unanswered, which, given the high cost of carrying out such studies and the long time taken to obtain the results—a further cost to the industry, was surprising. At the same time the large cosmetics companies began to realize that they had, over the previous years, developed a large database containing not only the results of animal studies but also human trials and patch test results and were able to relate these data to marketplace performance. Therefore, they were able to relate what happened in real life to the sort of effects that are seen in limited user trials, could see how user trials relate to patch test and other human experimental information, and relate what happens in experiments on people to what was found in animals. They also of course knew exactly what was in each formulation they tested and made.

What was more important in terms of assessing the safety of their products was that they had staff who had themselves done various aspects of the work, had access to all the data, and who had experience and expertise that other, less conscientious manufacturers and in particular the regulators and law makers did not possess.

The outcome of this combination of political pressure by the animal rights activists, sound knowledge and experience in the hands of a few major companies, and a desire by the regulators to be seen to be doing the "right thing", particularly in the UK (a nation that has a Royal Society for the Prevention of Cruelty to Animals (RSPCA) founded in 1824, but only a National Society for the Protection of Cruelty to Children (NSPCC) founded 60 years later in 1884), was that a political decision to ban animal testing of cosmetics was relatively easy to take.

The EU began the regulation of cosmetics in 1976 with the publication of the first Directive[6] and this was (and still is) amended regularly. The publication of the 6th Amendment, which came into force in 1996, requiring the professional assessment of any cosmetic product that was placed on the market within the EU was a seminal moment in the testing of cosmetics on animals. The UK regulatory authorities had increased the controls of all animal experimentation and took the implementation of the 6th Amendment as an opportunity to ban completely the testing of finished cosmetic products on animals. This ban, coming into force in 1996, was unique to the UK. Other countries, notably France and the USA still continued to assure safety of cosmetics by recourse to animal studies.

The 7th Amendment to the EU cosmetics regulations introduced the ban that had been operating in the UK to all countries of the EU and changed the assessment process by banning the use of animals in the testing of cosmetic products and cosmetic ingredients for any product sold within the EU. This was contested by France in the European Court, but their objection was dismissed in 2006. The UK model, effective since 1996 is now applicable to all EU countries.

What then is the benefit of this ban and how are safety assessments of cosmetics to be made when the use of animal studies is no longer allowed?

1.3 The Risk Assessment Process

The safety assessment of cosmetics is a risk-based process. Risk is a product of the intrinsic properties of a material and the exposure. Nothing is risk free; but if materials of low intrinsic health hazard are employed, then risks will be low, and if exposure is limited by either amount in contact or time of exposure, then again risks can be low. Risk is not a simple linear relationship between hazard and exposure, as demonstrated in Figure 1.2.

As mentioned above, nothing is risk free and this is true of cosmetic products. For the majority of people, most products can be used with no adverse

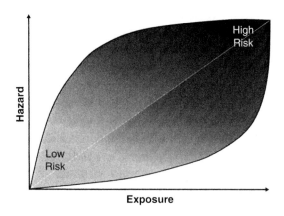

Figure 1.2 Relationship between risk, hazard, and exposure.

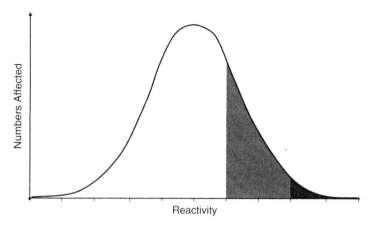

Figure 1.3 Difference in number of people affected if susceptibility is doubled (more people are reactive).

effects at all. Experience has shown that the response of the human population to most materials follows a classic bell shaped curve as shown in Figure 1.3.

The number of people affected adversely by any stimulus is shown on the vertical axis and the reactivity is shown on the horizontal axis. There are a few people for instance who will not react to any but the most severe stimulus, these are shown at the left hand "tail" of the curve; as reactivity increases, the numbers increase and as people become more and more susceptible the numbers decrease until at the right hand "tail" you are left with those unfortunate individuals who react adversely to things that the majority of people can live with quite happily. The total number affected is shown in the area under the curve. In the examples shown, the dark gray portion represents, on a scale of 1–10, those individuals who are level 8 and over in terms of their reactivity. As you can see, the total number of people as given by the dark gray area is quite small in relation to the total. If, however, the degree of reactivity is doubled as, for instance, by doubling the concentration of a particular ingredient, or increasing the amount applied, then as shown by the light gray portion of the diagram, the number of people who may be adversely affected is increased significantly. This phenomenon, typical of all biological systems, is the reason why concentration is so important in the safety assessment process. Concentration affects not only the scale of a reaction in any individual but also the number of individuals who will respond in the population.

1.3.1 Regulations

The regulatory environment in Europe, which has proscribed the testing of cosmetics in animals, also contains lists of materials that may and may not be used in cosmetics. These are listed in the schedules to the regulations and comprise the following:

- Schedule 3: prohibited substances—769 cosmetic ingredients and 34 fragrance ingredients
- Schedule 4: 158 restricted substances—concentration limits and label warnings required
- Schedule 5: permitted colors (and restrictions on use if necessary)
- Schedule 6: permitted preservatives (and limits on use)
- Schedule 7: permitted UV filters (and limits on use)

These lists are reviewed on a regular basis by a group of experts from across Europe, known as the Scientific Committee on Consumer Products (SCCP),[7] who ensure that the information available on each ingredient is adequate to enable an assessment of safety to be made. Irrespective of this requirement on composition, there is the umbrella requirement that each product is safe for use: it is perfectly possible that a formulation that is in full compliance with the compositional requirements of the law is still, in the opinion of the assessor, not safe for use and must therefore be rejected.

1.3.2 Role of the Safety Assessor

In European law the safety assessor has two distinct roles and as a professional consultant, a duty to their client, that is,

1. To ensure legal compliance: The assessor is legally liable for the assessment.
2. To protect the consumer: The assessor is an integral part of the legal process and is charged by law with protecting the consumer.
3. To protect the manufacturer: Product developers are concerned with new and marketable products and may not always take sufficient account of the health risks associated with the formulation. The assessor may be the last line of defense against product liability lawsuits.

The EU Regulations specify the qualifications and experience that are required to carry out an assessment; it is also stipulated that the assessor must have gained that experience within the EU. Qualifications required are: Doctor of Medicine (+3 years cosmetic safety work), Qualified Pharmacist (+3 years cosmetic safety experience), Chartered Chemist or Chartered Biologist (each of which requires a minimum of 3 years experience to achieve chartered status). A qualification that is becoming increasingly accepted across Europe is registration as a EUROTOX Registered Toxicologist.

The greater experience that an assessor has of the assessment process, the greater reassurance of safety and the less uncertainty that there is of the outcome of the assessment. This mirrors the experience of the development program in that the longer the product has been sold and the higher the number of people who have been exposed, the greater reassurance that it is safe for use. This is shown diagrammatically in Figure 1.4.

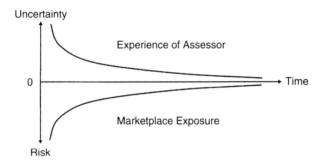

Figure 1.4 Assessment and risk diminish with expertise and sales.

1.3.3 Guidance

The EU Scientific Committee for Cosmetics and Non-Food Products (SCCNFP), now reconstituted as the Scientific Committee for Cosmetic Products, has published guidelines for the assessment process for cosmetic ingredients.[8] Also included in the document are suggestions related to the assessment of the safety of the finished product; however, this is restricted to five short paragraphs on pages 79–80 of the document. Overall, the recommendation is that the toxicity of the ingredients, the physico-chemical nature of the product, the microbiological status of the product, and the way in which it will be used must be taken into account (i.e., risk = hazard × exposure).

1.3.4 Safety Assessment

The first step in the assessment process is to establish the physical nature of the product and how it will be used. This means obtaining information on the physical nature of the product and asking questions like:

- Is the product liquid, solid, or gaseous? If liquid, what is the viscosity? Is it water or solvent based? If water based, what is the pH? If solvent based, is it flammable? If the product is solid, is it particulate and inhalable? If particulate, what shape and size are the particles? Does it pose a risk to the lungs, the eyes, or to the sensitive genital tissues? If the product is gaseous, is it from an aerosol under pressure, a trigger spray, or a button spray where volumes and particles sizes can all be significantly different? If an aerosol, what is the propellant?

The physical nature of the product determines how it can come into contact with the human body and thus describes one aspect of exposure.

The second step in the assessment process is to determine how the product will be used. Will the product be applied to the skin and is ingestion or eye contact possible? Will the product be applied in the morning and left on all day? Will exposure be brief, such as with a shower gel? For example, lipsticks and lip products will be ingested, hairsprays will be inhaled, and skin lotions will be applied repeatedly.

Stage 3 is to make sure that the formulation is legally compliant. Are all of the ingredients allowed and are they all within their concentration limits? Are there any warnings that are required?

Next the assessor must look at the information available on the ingredients. Are there any ingredients that pose particular risks? Some plant extracts are known to have potentially adverse effects; for example, rosemary is known to induce abortion. What, in particular, are the allergens present in fragrances, essential oils, and plant extracts. Are they present at a concentration that requires labeling? And if the product is to be left on the skin, what are the dangers of eliciting an allergic reaction, or worse, inducing allergy in a previously healthy individual? In all of this the assessor must bear in mind the variability of individuals, and acknowledge that it is not possible to give assurance of a risk-free product; only one that will pose an acceptable risk to the great majority of users.

With all cosmetics there are levels of customer complaint that are considered "acceptable." This varies with product type—a level of adverse reaction that is considered acceptable for a depilatory cream would, for example, be a cause of great concern if it resulted from use of a face cream. All major retailers and manufacturers have their own rule of thumb ratios that they use for assessment purposes. For example, if one complaint is received for every 50,000 units sold, that may be acceptable, but if the rate increases to 1:40,000 then there may be a need to monitor and begin investigation, at 1:30,000 a serious investigation is necessary and below 1:20,000 the consideration to withdraw the product will be at the forefront for the manufacturer or retailer. If the product is a depilatory cream, however, the numbers may be an order of magnitude lower and recall may be considered if the complaint level increases beyond 1:2,000 units sold.

The marketplace is the final arbiter of what is safe. As discussed in the "Introduction," perception of risk and individual control are important in determining what is acceptable and what is not. Only when a product has been sold for many years with no adverse effects reported can you be absolutely sure that it is safe, and even then there is always the possibility of a new customer coming along who just happens to be allergic to one particular ingredient or combination in your formulation!

1.3.5 Assessment Contents

The final assessment should comment upon and give the rationale behind any comments on the effects of the product both as sold and when used on:

1. The skin: Irritation, sensitization potential, dermal toxicity, and potential phototoxic effects should be considered.
2. The eye: Eye irritation of the product as supplied and when in use may be significantly different. For example, a bubble bath straight from the bottle may present a significant risk of eye irritation, but when diluted for use in the bath water, it will be fairly innocuous.
3. Ingestion: Again a bath gel if swallowed could be very unpleasant, but swallowing bathwater, as far as the bath gel is concerned, should be fairly uneventful.
4. Inhalation is unlikely for most products; however, any product that is sprayed will present an inhalation risk, as

will many powder products. Also consider the effects of vapors—for example, applying nail polish in a small car could lead to drowsiness and potentially a fatal accident if driving. This is a foreseeable use of nail polish and should be considered as a part of the assessment.

5. Finally, there should be an overall toxicological assessment. There may be a risk with the bath gel, but the risks can be reduced if people are warned and the product is not used as a plaything by children. The actual overall risks to health when the product is being used is the final factor that decides whether a product is safe and can be sold, or whether they are unacceptable and the product is rejected.

1.3.6 Other Tests

Some assessors and companies do not consider that an assessment can be complete without some test data. As animal studies are no longer an option, recourse to human patch tests or possibly cytology tests are often called for.

We do not believe, however, that there is much to be gained from such activities. An experienced assessor should be able to gauge from the information available on the ingredients whether a product will be an irritant to the eye with far greater accuracy than the results obtained on a particular cell line. In the same way, data derived from single or repeated patch tests are of limited relevance in assessing irritancy in use and of no value at all in assessing sensitization potential.

You cannot of course carry out a sensitization test in man: For the issue that is the prime cause of concern in all cosmetic safety assessments—sensitization—there are no cytology tests, animal tests are prohibited, and human tests are unethical. Assessment by experienced personnel, using data on ingredients is the only way to address the major issue of safety assessment. It seems perverse to require testing for effects that are far easier to judge.

1.4 Conclusion

The assessment of cosmetics by qualified and experienced personnel is the only way that the law allows the safety of cosmetics in Europe to be determined. In the UK, experience over the past 12 years, drawing on the expertise gained through assessing a large number of different formulations, this process can be both quick and effective and above all ensures the safety of the consumer in the way that the law intends. I would also contend that the assessment process draws on the sum of all that has gone before and represents a considerably larger base of knowledge than any other assessment process, particularly when many of the products that constitute the knowledge base have been used by millions of people around the world. This is represented diagrammatically in Figure 1.5.

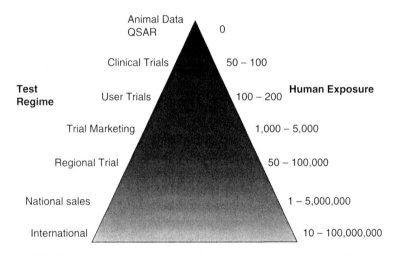

Figure 1.5 Human exposure and degree of safety reassurance increases as more people are exposed over time.

In general an experienced assessor in the EU will reject 10–20% of all formulations submitted. Providing that the manufacturers and formulators involve the safety assessors at the outset of the development process, and not in the week before the product hits the shelves, the safety assessment can be a cost-effective part of the product development program, and the safety of the consumers will be protected far more effectively than using animal or other models where a crude numerical value of toxicity is the only result available.

References

1. Ames, B. N. (2006) Putting Cancer Risks in Perspective. Paper presented at the HBA Regulatory Forum, New York, 6 September 2006, by Prof. B. N. Ames, Children's Hospital, Oakland Research Institute, University of California, Berkeley, CA.

2. Mortality Statistics: Cause (2006), Review of the Registrar General on deaths by cause, sex and age in England and Wales, 2005. Her Majesty's Stationary Office, London. ISBN (10) 1-85774-641-4. Available online at http://www. statistics.gov.uk/downloads/theme_health/Dh2_32/DH2_No32_2005.pdf
3. Council Directive of 27 July 1976 on the approximation of the laws, regulations and administrative provisions of the Member States relating to restrictions on the marketing and use of certain dangerous substances and preparations (76/769/ EEC). OJEC L 262 27.9.1976, p. 201.
4. International Agency for Research on Cancer, Lyon. Available online at http:// www.iarc.fr/
5. The Dangerous Substances and Preparations (Safety) Regulations 2006. Statutory Instrument 2006 No. 2916, 6 November 2006.
6. Council Directive of 27 July 1976 on the approximation of the laws of the Member States relating to cosmetic products (76/768/EEC). OJEC L 262 27.9.1976, p. 169.
7. Scientific Committee on Consumer Products (2007) Information on activities, publications, membership, etc. Available online at http://ec.europa.eu/health/ ph_risk/committees/04_sccp/04_sccp_en.htm
8. EU (2003) The SCCNFP's notes of guidance for the testing of cosmetic ingredients and their safety evaluation, 5th Revision. Available online at http:// ec.europa.eu/health/ph_risk/committees/sccp/documents/out242_en.pdf

2

Regulatory Developments in Canada, Japan, Australia, China, and India*

Janet Winter Blaschke

*International Cosmetics and Regulatory Specialists,
LLC, Manhattan Beach, CA, USA*

2.1 Introduction

In the area of regulatory affairs, myths are prevalent, and it is important to have current information about the changing regulatory scene. This involves vastly different areas of the world, and a brief summary herein should give clarity to a few of the diverse issues that exist. Consider that any country that needs a passport to enter is one that has a different regulatory system than the US. There are 193 countries in the world, and none of them have an identical regulatory system. With regard to Cosmetic Legislation, some countries require registration of products with the regulator, others do not. All countries disallow some ingredients, but of course not the same ingredients. In some countries it is not possible to even import or sell the product directly as an American company; based on government rules, a local based company may be necessary.

---

*Based on a presentation made at the Health and Beauty Aids Regulatory Symposium in September 2006.

C. I. Betton (ed.), Global Regulatory Issues for the Cosmetics Industry Vol. 1, 21–33
© 2007 William Andrew Inc.

In the sections below, I will outline the regulatory system in some of the major export markets of Canada and Australasia. The European Union is covered elsewhere in this book and I will leave that to others. I also address some of the requirements applicable within the US; the state of California for instance, has its own requirements in addition to the general Federal Laws that apply to all.

2.2 Canada

Though Canada is closely associated with the US, it has its own identity in the laws which apply to cosmetics and cosmetic-type products. For many years our neighbor was fairly quiet. Now things have become interesting. There are two big, recent changes that are very important to anyone who does business in Canada.

Until very recently, contrary to popular belief, there was no law that required an ingredient list on cosmetics in Canada. As of November 16, 2006, those days are gone. It is now law to have an ingredient list on cosmetic products marketed anywhere in Canada.

The biggest problem historically in Canada was the bilingual language requirement. American products that were imported into Canada often had the ingredient lists on the package, as required by US law. However, even though the ingredient list was not required in Canada, the very fact that the ingredient list was on the product in English, triggered the requirement for French. The ingredient list actually turned out to be two ingredient lists: one in English and other in French. Needless to say that made for a very crowded label.

The general rule for Canada with a few exceptions is "if it is in English, it has to be in French." Many American companies large and small feel the pain of enforcement of the dual language requirement by the Quebec Language Ministry. French cannot be of lesser predominance than English with very few exceptions.

But there is actually some good news for those who want to have a product that can be labeled for both the US and Canada. Required ingredient labeling which is now in force may not sound like good news. However, it avoids the most of the requirements for an English and French ingredient list. The Canadian government now recognizes most of the standard ingredient terms. Of course, it would be too easy if the two systems were identical. Laws just do not work like that—it seems to be some kind of rule of the Universe. However, this allows companies to make one ingredient listing, if carefully calculated. The mine fields are some of the ingredients which are NOT the same as the US. In those cases, the bilingual requirement often applies.

Health Canada, the agency which regulates cosmetics, has stated that there is no grace period after November 16, 2006. The rulemaking allowed for a two-year notice prior to enforcement. The Canadian Cosmetic, Toiletry, and Fragrance Association (CCTFA) did an excellent job in negotiating the two-year phase-in period. This is the first major labeling change in Canada for several years. Companies that sell products in Canada must actively deal with these changes if they have not done so already.

There is still some confusion about this requirement and its enforcement. French is the official language in the Quebec province, whereas English is the official language in the other Canadian provinces. Mandatory labeling requirements must be in both languages, such as warning statements. The French language Ministry dictates also that "any inscription on a product" appear in French. Other languages may appear, but no other language can predominate over French. So on a practical level, as well as an enforcement level, product sold in the province of Quebec must have a completely bilingual label.

There seems to be little confusion on the issue of enforcement. Authorities in the province of Quebec regularly check the shelves of the stores and do not hesitate to cite a company for noncompliance. Product distribution can be halted and companies are assessed fines for not complying with the language requirement. This happens on a regular basis across all industries, not just cosmetics.

Some companies try to avoid the requirement with the thoughts of selling product only outside Quebec. This is not something we typically recommend. Two of Canada's largest cities, Montreal and Quebec City, are in this province. Many chains of distribution, both large and small, will have inlets to Quebec unbeknownst to the original product owner. Once the product appears on the shelf and is enforced, the product will need relabeling to appear on the shelf, and again, fines can be levied. It is simply more practical to adhere to this requirement so that the product can move freely throughout Canada.

The second recent, drastic change in Canada is the creation of the Natural Health Products (NHP) program. This regulatory system was originally created to have a category for items with natural ingredients that had a history of therapeutic or practical use, primarily dietary supplements or ingestible products, and any products that have had a long history of homeopathic use. This new category did not exempt cosmetics, as many had hoped.

This is something that many government agencies have wrestled with for over 10 years. How can you assure that natural products are safe, and also effective, so that they are legitimate for consumers? It sounds easy enough, but for scientists it is a large dilemma. Most natural products that have been used for many years cannot be scientifically quantified. A simple herbal extract can actually contain hundreds of chemicals, and some are not even defined yet.

This new category casually called NHPs contains some products that used to be considered as cosmetics.

As an industry, we have been aware that there is always a group of products which have been considered to be more therapeutic products, but without requiring a prescription. This group of products was called Category IV drugs, and included things like sunscreens, acne products, and fluoride toothpaste. With the change to NHPs, some but not all of these Category IV products have been changed.

For instance, a sunscreen that contains only titanium dioxide would now be considered an NHP. However, if you have other sunscreens like Padimate O, or other organic sunscreens with titanium dioxide, it would be considered under the existing Category IV law and would require a drug approval. Those of us who get products approved for a living, now concede that the once dreaded drug approval is often preferable to the NHP process.

An NHP approval requires an in-country representative which is licensed specifically for this purpose. This representative must have an official location in Canada with personnel present and cannot simply be a post office box. The submissions to Health Canada are quite lengthy and require specific quality standards. This approval process from date of submission takes approximately 15 months at the time of this writing.

So where there were two categories previously: cosmetics and Category IV drugs, now there are three: cosmetics Category IV drugs and NHPs. Products commonly sold in our industry fit in one of these three categories.

After many years of mostly status quo in Canada, companies now have had to revisit all of their products to determine which have changed.

There is only one best way to deal with this, and that is to work with formulators, marketers, and regulatory professionals. Is it worth having a Chamomile lotion with specific claims, to wait for an NHP approval? Or can it be a lotion with similar characteristics that can be marketed without the delay of the approval?

2.3 Japan

The first country that I dealt with in Corporate Regulatory Affairs was Japan. This is how many people have started their regulatory careers. This often has occurred because Japan is one of the first truly foreign markets to which American companies expand. There often has been the first realization for companies that Dorothy and Toto are not in Kansas any more.

There is a story/urban legend that tell a lot about the Japanese quality standards and also cultural misunderstandings. They are still laughing about this at IBM. Apparently, the computer giant decided to have some parts manufactured in Japan as a trial project. In the specifications, they set out that they would accept a deficiency level of three defective parts per 10,000 units.

When the delivery came in there was an accompanying letter. "We had a hard time understanding North American business practices. But the three defective parts per 10,000 have been separately manufactured and have been included in the consignment. We hope this pleases you."

As a regulatory consultant I am pleased to say that the last few years there have been truly earthshaking changes at the Ministry of Health, Labor and Welfare (MHLW) in Japan. Their regulatory system has had what can only be described as a major overhaul. The net result has been redistribution in the responsibility of marketing products in Japan.

When I was first obtaining approvals for hair care products in Japan, there was a mountain of data required. Once the mountain of data was compiled and submitted, the approval time was a minimum of six months. There was a formal approval that was issued, along with a number that was assigned to the product. This resulted in great delays in bringing products to market.

After the changes in the last few years, there is no longer a requirement to have a formal approval for products that fall under the definition of cosmetics. This process has changed to that of notification. The notification simply advises the government of the product and to whom the responsibility lies, and there is no waiting period of 6 months for a formal approval. This became effective on April 1, 2001.

However, there is no free lunch. Again, similarly to Canada, when the approval process was downgraded to a notification, the Japanese government felt that there should be more controls in place. The Japanese MHLW wants to make sure that it protects the health of its citizens. In the lengthy approval process, the Government primarily assumed responsibility for product safety. Without the lengthy review process, the Japanese distributor now has the responsibility of ultimate product safety. This requires that a technical professional be on staff. The net result to American companies is that their distributors may be asking for more information, and may be becoming more of a partner in the process, since the responsibility is theirs. All importers and manufacturers in Japan must be licensed to do so. An inspection must be passed in order to get a license.

Another trade-off for the elimination of formal approval and safety review is that of more communication to the consumer via the label.

Labeling is something that is continuing to change. As with Canada, there previously has been no requirement for a complete ingredient list on cosmetic products. The previous system required an ingredient list with only a few ingredients which were thought to be more chemically reactive than others. Now the ingredient labeling must include all ingredients used in batching a cosmetic product. Each ingredient must be listed by its Japanese name. Ingredient names in English do not fulfill this requirement. Ingredients used on labeling must apply for a Japanese name for this purpose.

Pharmaceutical Affairs Law may sound like it only pertains to prescription drugs. In fact, it covers cosmetics and quasi-drugs as well.

A quasi-drug is something that has a "mild effect on the body" and has a definite "purpose of use." This includes such things as certain kinds of skin lotions, antidandruff shampoo and acne products. quasi-drugs must still get a pre-market approval prior to sale.

The ultimate effect on American companies is a closer relationship with their Japanese distributor and/or the Japanese manufacturer, and increased labeling requirements, in exchange for a faster time to market.

Ultimately in Japan as with any other countries, cultural issues are a part of all aspects of life. This includes marketing products. The authorities are very serious about public health. There is also a prevailing belief in Japan that Japanese skin may be more sensitive to some chemicals and ultimately, some products. For many years, it was thought that this was simply a reason to create a roadblock to new chemical approvals for use in cosmetics. However, there has been some recent scientific evidence that effects of products in fact can be different between races, however politically incorrect that may be interpreted. In the end, it is simply Public Health and safety that is of primary focus.

Enforcement focuses on ingredients as well as product categories. The Ministry still regularly tests products for ingredients that it claims to have in the product, as well as testing for other ingredients that are not claimed in the products. It is important to be very accurate in listing ingredients, especially concerning colorants, preservatives and sunscreens.

Cosmetics and quasi-drugs are both regulated under the same umbrella in many respects. However, quasi-drugs are required to be made to a higher standard and are subject not just to pre-market approval, but special testing as well. It is of utmost importance that companies know the category to which the product belongs, and how enforcement is handled. There is much for the licensed importer/distributor to lose with regulatory challenges. In addition, their license is nullified, a new importer would have to be located, and this is the responsibility of the product owner, such as an American company.

I have heard many times that Japan is most difficult market for regulatory challenges. This fallacy may have its roots in the 1980s when there were many restrictions on ingredients, and the products were subject to formal approval.

While Japan is still not the easiest market for products, it is certainly not as challenging as other markets, such as Korea.

We should be grateful for the fact that there are many, many more ingredients allowed in Japan than in the past, but this market should be cautiously approached. It is important to understand that products are tested and there are still restrictions on some commonly used ingredients.

With any large change, such as the shift of responsibility for product safety, we can expect that the Pharmaceutical Affairs Law will continue to evolve over time. Industry should continue to monitor changes and continue their participation with the Japanese MHLW through industry groups.

2.4 Australia

Australia also has its own unique regulatory system. Again, it has several categories for cosmetic-type products. The first and most obvious is that of "cosmetics." Another category is "therapeutic products" which is similar to Canada's Category IV drugs, and the US over-the-counter drugs.

The classification can be determined by two parameters:

1. Proposed use
2. Composition of the product

The proposed use of the product takes into account any labeling claims (which can include the product name) product inserts and advertisements.

The composition of the product of course involves the ingredients. A particular ingredient or a specific level of an ingredient can cause the product to be a therapeutic good and not a cosmetic.

The listed medicines category, also known as listed therapeutic goods, is somewhat based on the considered risk of the product. Products over SPF 4 are regulated as listed medicines or listed therapeutic goods. SPF products of 4 or less containing human or animal tissue are regulated as well. This is evidence of the risk concern of animal or human derived ingredients. Products are subject to a formal approval process, less arduous than prescription products. One of the complications of this category is that only previously approved ingredients can be used in listed products. These ingredients must have been reviewed for safety and efficacy.

Exempt therapeutic goods can make certain claims, while not requiring listing. The category is small and specific and includes antiperspirants, antiacne cleansers, antidandruff products, and fluoride toothpaste. Product claims and ingredients are primary in the determination of the category for each product. If you think that the product may be "listable," it probably is. Labeling does vary between categories, and can also be different within a category. Cosmetics do not require a registration number on the label, whereas the registration number for therapeutic goods must be shown.

Surprisingly, Australia has its own approved ingredient list for cosmetics and chemicals used in other products. Australia maintains the Australian Inventory of Chemical Substances (AICS). If an ingredient is to be used in a cosmetic, it must be listed in AICS. If it is not, it must be submitted under the National Industrial Chemicals Notification and Assessment Scheme (NICNAS). If the ingredient is approved, it is allowed to be used in a cosmetic product imported or manufactured in Australia.

So even though there is no official approval for cosmetic products, the raw materials must all have an approval in order to be used in a marketed product. There is no special labeling for NICNAS approval.

Therapeutic goods must also use only approved ingredients. The goods must be included in the Australian Register of Therapeutic Goods (ARTG). Safety and efficacy and good manufacturing practices data must also be included. The time for approval of therapeutic goods is approximately 3–6 weeks.

2.5 China

If you pick up any business magazine or newspaper you will always find a reference to China and its economic changes. Here are some statistics.

China has a population of 1.3 billion people, the most populous country in the world. The Chinese middle class consists of 250 million people. That is, only 16% less than the current population of the US of all classes. In one year, disposable income rose almost 12%. At the same time, spending on consumer products increased almost 14%.

So the rumor that it is worth going through Regulatory hurdles might just be true. And challenging they are . . .

As China works toward being an even bigger world economic power, they are faced with a dilemma—how to employ its citizens and foster the economic growth. Of course, this is a matter of dollars and cents. There is a clear disincentive to allow foreign products to be imported into China, which is one of the basic principles at work in other countries as well.

While I am not suggesting that this is a clear reason for regulations, it is certainly a real-life consideration. This perspective can be helpful in understanding some of the logistical challenges in breaking into the Chinese market.

It is possible that American companies have historically somewhat contributed to the need for regulations. There have been instances where American company has had product that does not live up to its American quality standards and have been known to sell product to another country. The regulations which review quality can keep this from happening.

However, there is a real help in working with the government. If there is a Chinese company which imports a product, they can be a tremendous help in working with the government. Even better, sometimes bulk product can be imported and filled locally. Again, the support of the local industry can go a long way. The best alternative in regulatory terms is to have as much local handling as possible done locally.

While this is a good regulatory option, there are other things to consider based on first hand experience. I was responsible for setting up a laboratory in a factory in China. Many things that we take for granted are either nonexistent or are complete luxuries. One example is consistent electrical power for manufacturing. Another is computer systems that can trace inventory. So if the choice is made to manufacture locally, the infrastructure should be thoroughly investigated and considered seriously.

Another interesting part of the Chinese registration process is the sending of samples to the government for testing. Typically, 15–100 actual product samples are required to be tested prior to product approval. While this may rankle some companies, it is generally easiest to submit samples.

BSE (bovine spongiform encephalopathy) is known as Mad Cow Disease. There have been a number of cases discovered, originally in the UK, then other countries including three cases discovered in cows in the US. While the science involved in this type of disease is not well developed, there is a concern on the part of some governments to assure that there is no possibility of BSE in products. There has never been a scientific link between BSE and dermal exposure in humans.

However, to allay concerns, the Chinese government has required specific documentation for products coming from countries which have had even one case of BSE. This of course includes the US. China has recognized only a certificate of compliance from the Health Agency or another body that it agrees can issue the certificates. In the case of the US, this was the Cosmetic, Toiletry, and Fragrance Association in Washington, DC. Companies were required to certify that their products do not contain BSE-suspect materials. However, very recently China has abolished this requirement for new product submissions. It is proof that regulatory requirements really can change for the better.

In spite of some of the hurdles in penetrating the Chinese market; it is often worthwhile, since American and European goods are potent status symbols in the new Chinese economy.

There are many other countries that deserve some attention, and others will be briefly discussed below.

2.6 India

India has the largest population in the world, after China. The population is almost 1.1 billion, to China's 1.3 billion. The Indian market has just begun to allow foreign products legitimately in their market. While there is still a long way to go to enter the market, this bears close examination for expanding businesses. The regulatory system is still evolving, and appears to be going from impossible to may be possible. Interested companies should follow developments closely.

2.7 Trade Alliances

Slow progress continues to be made in trade alliances such as Association of Southeast Asian Nations (ASEAN) in Asia and MERCOSUR (Southern Common Market) involving some South American countries. It is hoped that these trade alliances over some time, will smooth the way for some harmonized regulations in groups of countries. So companies should consider donning rose colored glasses for a few years.

The current member countries of ASEAN are Brunei, Cambodia, Indonesia, Laos, Malaysia, Myanmar, Philippines, Singapore, Thailand, and Vietnam.

MERCOSUR is made up of five member countries: Brazil, Argentina, Paraguay, Uruguay and Venezuela. Other associate and observer members include Bolivia, Chile, Colombia, Ecuador Peru, and Mexico.

At this time, it is not advisable to expect one harmonized system between all of the countries in a trading bloc. However, these alliances, hopefully some time in our lifetime, may help smooth the way for a degree of alignment.

2.8 California

The state of California is the sixth largest economy in the world and is larger than many countries in gross domestic product. California has learned to flex its muscles in determining its own unique regulations which are in force in addition to US Federal regulations. Familiar topics include the regulations for volatile organic compounds (which can contribute to air quality) and the Senate Bill IV8IV (which requires registrations for any ingredients contained in products that are defined as carcinogens, referencing several government lists). Companies should be familiar with California's regulations, since the State is very active in enforcement.

2.9 Conclusion

International expansion is challenging at best, but if good planning and analysis is done, it may be possible to:

1. Combine labeling for more than one country wherever possible.
2. Assure that formulations and their ingredients will be acceptable in targeted countries.
3. Forecast on an ongoing basis so that timelines for approvals and formula modifications can be anticipated in advance. It is best to know the show stoppers before a new distributor has placed an order!
4. Fine tune intracompany communication, which is essential for complex International expansion, whether for a small company or a large one.

Of course, the world continues to shrink. It always amazes me how much government agencies half a world away know about budding regulations in the opposite corner of the world. California's VOC regulations are being replicated in many countries, and some government officials overseas are better versed than some American companies.

As with anything else in life, the Internet plays an increasingly important role. Government agencies can communicate with each other very rapidly, especially where public health is concerned. Problems with one government health office in the US can be under scrutiny in Canada very quickly.

There is no other way to prepare than to monitor changes as they begin. Many companies have been blindsided by Europe's new chemical regulations, though they have been underway for several years. It pays to commit completely to foreseeing the requirements of International Business and their Regulatory commitment.

3

The REACH Regulation of the European Union

C. I. Betton

Delphic HSE Solutions Ltd
Camberley, UK

3.1 Introduction

The origins of this chapter, like the book itself, lie within a paper presented at the first Regulatory Summit at the Health and Beauty America (HBA) Exhibition and Conference held in New York in September 2006. Unlike most of the other chapters, however, when the presentation was delivered, the legislation that was being talked about was not yet finalized and there was an element of speculation, not just on what the final format would be, but also whether the whole tranche of the REACH (Registration, Evaluation, Authorisation of Chemicals) legislation would ever find its way onto the statute books. Agreement was finally reached, however, on the night of Thursday, November 30, 2006, on the most radical piece of legislation ever to be passed concerning the manufacture sale and use of chemical substances, not only within the European Union (EU) but also anywhere in the world. The final wording was passed by the Parliament and Council on December 18 and published in the *Official Journal of the European Union* on December 30, 2006, coming into law on June 1, 2007.[1]

C. I. Betton (ed.), Global Regulatory Issues for the Cosmetics Industry Vol. 1, 35–47
© 2007 William Andrew Inc.

REACH fundamentally alters the way in which chemicals are regulated and, recognizing the global nature of the regulatory process, may well prove to be a taste of things to come for all manufacturers of chemicals, wherever they may be situated: already China is looking at the possibility of introducing similar laws.

In this chapter, I give some background to the REACH regulation, explaining its evolution and purpose. I also explain how REACH is expected to work, this is of necessity speculative, as the first thing that the regulation does is to establish the new European Chemicals Agency which will administer and run REACH, and until this new agency starts work in June 2008, no-one will know for certain how things will work in practice.

Finally, I address the registration process, what the consequences will be for businesses and express some of the potential concerns that still exist regarding the introduction of REACH, particularly for the cosmetics industry.

REACH is radical but it is also very simple. The fundamental principle underlying REACH is that if a company manufactures chemicals in the EU or if anyone wants to import chemicals into the EU, then they must know what those chemicals are used for, what the effects on man and the environment will be and must provide an assessment of risk to the authorities to demonstrate that any risks are known, understood and manageable. In short, they must take responsibility for their actions.

Behind that simple idea is a piece of legislation that is over 1,300 pages long, covers standard methods for determining physico-chemical, toxicological and environmental properties, and replaces over 40 existing pieces of legislation.

The good news is that cosmetics, being subject to a risk-based health assessment before they can be marketed in the EU are exempt from REACH *but only as far as health effects are concerned*. The environmental aspects of all cosmetic products are subject to REACH and an environmental risk assessment will be required for cosmetic ingredients where the total annual sales are more than one tonne per year. Packaging will also be subject to the requirements of REACH.

3.2 Why REACH?

In 1981, the EU introduced legislation that made it mandatory that all "new" chemicals placed on the market should have a dossier of physico-chemical and toxicological data including some rudimentary ecotoxicity information, that would be submitted to the competent authority in each member state. These data were based on tonnage sold and the volume, complexity and cost of information needed increased as tonnage increased. Since 1981 approximately 1,300 chemicals have been introduced and accompanied by the appropriate data submission, have been marketed. Prior to the introduction of this legislation, the chemical industry was invited to register all existing chemicals with the EU: How could a chemical be said to be "new" unless there was a definitive list of "existing" substances? Seeing the legislation on new chemicals on the horizon, the chemical industry registered every chemical that had ever been sold, synthesized or thought of, and even some that had not, reasoning that if they registered all chemicals then these were safe from a lot of expensive testing. Approximately 500,000 individual substances were originally registered.

The EU, however, not being completely naïve, then started to ask for all existing data on existing chemicals. This information, without assessment or analysis and certainly without concern over quality or good laboratory practice (GLP), was submitted in the form of a computer database disk, obtained from the commission and fed into the central data bank. Information requested, as well as health, safety and environmental data also included such basic information as how much was made of each compound. This exercise resulted in a database of about 300,000 industrial chemicals and is known as the European Inventory of Existing Chemical Substances (EINECS). New chemicals were placed on a new list—the European List of New Chemical Substances (ELINCS). Today, in order to be sold in the EU, a chemical must be on one of those two lists.

Having obtained the data that existed on existing substances the next phase was to perform a gap analysis and to identify substances of concern. The assessment criteria were based on tonnage, toxicity data—or lack thereof and read across from similar structures with known adverse effects. This resulted in a list of priority chemicals where further information was deemed necessary and the responsibility for working on these chemicals was apportioned to a number of the member states, each state taking the lead on a small number of compounds. This frankly is where the problems started, as the level of competence in the member states was variable, political considerations related to protection of home industries became clear and due to the way in which the EU operates constitutionally, the process rambled on for some years and nothing actually ever got done.

The existing situation was not sustainable. It is known that some chemicals placed on the market had adverse health effects—lead from additives in petrol adversely affecting children's learning ability, asbestos from insulation and automotive components leading to mesothelioma, vinyl chloride causing liver tumors in PVC workers and many more examples. While new chemicals were subject to investigation and scrutiny before they are sold, existing chemicals only revealed their adverse effects as a result of the illnesses and diseases that they caused. The situation was the same with regards environmental effects. DDT was banned as a result of its effects on raptor eggs, it had saved millions from malaria but the environmental consequences were considered unacceptable. Tributyl tin affected the reproduction of the dog whelk at the parts per billion level in seawater and so was banned for use in antifouling paints—except on warships! I hesitate to say it, but neither of these compounds would have been identified on the basis of the standard environmental screening studies employed then or probably now.

At the same time, public concern over health and environmental issues was running significantly ahead of the activities of politicians and a new way of dealing with the situation was needed. The process began with the chemicals white paper, a consultation document, published on February 27, 2001. Following extensive consultation the result is the REACH regulation, agreed on November 30, 2006, a mere five years and nine months later, a timescale that is breathtaking given the radical nature of the regulation and the complete reversal of responsibilities with regards chemical regulation that it is promoting.

3.3 REACH: Overview

The first thing to note is that REACH is a Regulation. The majority of EU legislation is in the form of Directives. A Directive is an instruction from the EU to the member states to implement certain pieces of legislation in their own Law, in their own way, but making sure that the spirit and intention as set out in the Directive is carried forward. This results in there being subtle but often significant differences in the way that each member state implements and enforces certain pieces of legislation. A Regulation is a piece of legislation that becomes law in each member state shortly after it is published in the *Official Journal of the European Union*. It becomes law as written with no further implementation or interpretation and it comes into force in all member states at the same time. A Regulation is truly a pan-European piece of legislation and REACH is a Regulation that came into force on June 1, 2007.

REACH is designed to replace over 40 existing EU Directives and much of the detail regarding the test methods for determining toxicity, and physico-chemical properties is unchanged. What is new, however, is the emphasis on risk-based assessments and the move away from a purely prescriptive form of data generation towards a more iterative process whereby if a scientific case for a particular course of action (or non-action) is justified then that can be pursued. There have been a number of pieces of legislation—cosmetics and biocides for example, where the emphasis has been more towards a professional assessment of the risks by industry and approval is given based on acceptance that the assessment may be checked and challenged but that it is still the industry assessment that is the fundamental step in the process. REACH places the responsibility firmly on the manufacturer to assess risk and to demonstrate that the chemicals made or imported can be used safely and that ultimate disposal will not pose health or environmental hazards.

3.4 REACH: Aims

There are two main and several subsidiary aims of the REACH regulation:

1. The first main aim is to transfer responsibility for the assessment of risk from the regulatory authorities to the manufacturer/importer in the EU.
2. The second main aim is to ensure that there is an adequate flow of information from the manufacturer/importer to enable the user of the chemical to handle and use the material in a way that minimizes risk and that information will flow up the supply chain to the manufacturer/importer so that they know how the chemicals are being used and can take account of those uses when assessing risk.

Following on from these two fundamental aims of the REACH regulation are a number of other ideals, some directly related, such as the substitution of active chemicals by alternatives of similar performance but lower toxicity and hence risk, to the formation of consortia of manufacturers, importers and users to share not just existing data, but also the costs of developing new information as required, thereby reducing the amount of animal testing that is carried out.

When a chemical is submitted for registration, there will be a number of requirements for information. These will be discussed in depth later on, but the submission for a chemical must include not only information of the properties of the chemical—factual data—but also an assessment of the risk and impact on health and the environment of the named uses of that chemical. It is estimated that for the majority of chemicals that there will be no further action required. For substances of very high concern, and there are estimated to be about 1,500 of these, then there will be requirements to substitute alternatives, if such alternatives are not available, then manufacturers will be required to carry out research to identify possible substitutes. As well as substitution, there will be restrictions imposed on the use of certain chemicals, based on the risk to users and the environment. This of course happens at the moment: lead-based paints cannot be used in children's toys or furniture; tributyl tin cannot be used in consumer antifouling paints; carcinogens cannot be supplied in consumer products, and so on.

These restrictions have up until now been applied at the end of the assessment process and are in effect retrospective. The future will be different and the restrictions will be applied before new chemicals enter the marketplace of the EU; application to existing chemicals is of necessity retrospective, but the emphasis for composition and scope of the risk assessment will originate with the manufacturer/importer.

In order to carry out any risk assessment, there is a need for information on the hazards associated with the material in question and these data, together with information on exposure, forms the basis of the risk assessment. One of the reasons that REACH was deemed necessary was that for may existing chemicals, there are few that have comprehensive toxicological data and even fewer for which environmental information is available. One of the aims of REACH is to ensure that these data gaps are filled *as far as it is necessary to do so*, in order for the risk assessment to be carried out. Generation of appropriate data is therefore a significant aim of the REACH regulation. However, it is also a stated aim of REACH that the amount of testing, and animal testing in particular, should be minimized.

These two apparently conflicting aims are being addressed in the first phase of the REACH regulation, as soon as the European Chemical Agency (ECA) becomes operational on June 1, 2008. The mechanism for achieving this end comes in two parts: first is pre-registration and second is the mandatory sharing of costs and effort in generating new information in the Substance Information Exchange Fora (SIEF) which are the first projected activities that will need to be undertaken under REACH. These are described below.

3.4.1 Pre-registration

REACH applies equally to new and existing chemicals. All are subject to a risk assessment before they can be legally sold within the EU, with responsibility falling on manufacturers and importers. For new chemicals the process is straightforward and a dossier must be submitted containing all of the relevant information (described later) as well as a risk assessment. However, for existing chemicals, the chemicals themselves are already on the market and in-use. It is acknowledged in the regulation that from a practical standpoint, it is not feasible to require that all registrations must be completed on the day that REACH comes into force, nor is it possible to remove chemicals from use until they have been registered. It would not be physically possible to generate and assess all of the data and it would cause economic collapse if 90+% of the chemicals in use in Europe were withdrawn overnight.

To overcome these difficulties the concept of pre-registration is built into the REACH regulation. If a manufacturer or importer pre-registers their chemicals with the ECA then they can continue to sell those chemicals within the EU. They then have a longer time span in which to complete the registration process under REACH. The timescale for the completion of registration is dependent on the properties of the substance and the annual tonnage that is either manufactured or imported into the EU. The timeline for pre-registration and registration is shown in Table 3.1.

The information that must be submitted at pre-registration, for all substances manufactured or imported into the EU in quantities greater than one tonne per year is simple and is as follows:

- Name of chemical substance
- EINECS number
- CAS number (or any other identity code if CAS and EINECS are not available)
- Name and address of registrant
    - Name of contact person
- Envisaged deadline date for registration
- Tonnage band

Table 3.1 Timeline for Registration Activities Required by REACH

Date	Activity	Tonnes per year
December 30, 2006	REACH regulation published in Official Journal	
June 1, 2007	REACH comes into force	
June 1, 2007	ECA formed	
June 1, 2008	Registration of new chemicals begins	
June 1, 2008	Start of pre-registration for existing chemicals	
December 1, 2008	End of pre-registration	
January 1, 2009	Publication of manufacture and import information on EU database	
February 1, 2009	Beginning of SIEF formation	
June 1, 2010	Registration deadline	
	High volume substances	>1,000
	Persistent, bioaccumulative, highly toxic to the environment	>100
	Carcinogens mutagens toxic for Reproduction, endocrine disrupters	>1
June 1, 2013	Registration deadline	
	Medium volume substances	>100
June 1, 2018	Registration deadline	
	Low volume substances	>1

This information will be submitted to the ECA by means of a software download that will be available free of charge from the ECA website. The information will be entered into a database, known as IUCLID 5 and which will form the basis of the entire registration process.

The importance of the pre-registration process and the 6 month window from June 1 to December 1, 2008 cannot be over emphasized.

If manufacturers and importers wish to make use of the extended deadlines for the completed registration process they *must* pre-register their substances; if they do not, and the window is missed, then the chemicals that they produce or import will be classed as "new" and will be subject to the full registration process *before they can be legally sold.* In short, they will have to be withdrawn from sale until a registration is completed.

3.4.2 Substance Information Exchange Fora (SIEF)

Assuming that manufacturers/importers wish to take advantage of the extended registration deadlines and have pre-registered their chemicals in the IUCLID 5 database, the ECA's next action is related to the aim set out in the REACH regulation of minimizing testing and the costs of testing and encouraging the submission of joint registrations for as many chemicals as possible and also minimizing the use of experimental animals. The SIEF is the first stage in that process.

The sharing of information and costs is not a new idea, industries have been generating data in response to regulatory requests in both Europe and the USA for many years. In the EU when the Commission requested existing data from industry on high volume chemicals in the early 1990s, as HEDSET submissions, it was possible to submit information either as individual companies or jointly, but not a mixture of the two. Naturally there were some hiccoughs, companies which had generated data—at cost to themselves—were reluctant to give this away in a combined submission without some financial recompense. Once they grasped the idea that they could submit the data on their own, but that they would have to complete the entire submission also on their own, the majority added their data to the general pot with often grudging good grace. The EU have learned from that experience and there is still the possibility that registration may be carried out by individual companies, providing that they can justify not taking part in a combined effort. The emphasis however is very much on co-operation. On January 1, 2009, one month after the closure of the pre-registration window, the ECA will publish on its website information on all of the chemicals that have been pre-registered and the contact details of the companies associated with each substance. The intention is that those with an interest in each chemical then get together to exchange information and compile the registration dossier, sharing costs equally and as all are involved in assessment of the existing data, minimizing the likelihood of duplicate studies.

There are some rules in the REACH regulation that cover the formation and functioning of a SIEF and the subsequent submission of registration documents. These will not be tested until January 2009 at the earliest so there is no indication of how effective these will be; however, the basic rules are shown below, as an indication of how data sharing is *supposed* to work:

- Data submitted to a SIEF shall be given freely and costs shall be shared in a manner agreed by all parties, if agreement cannot be reached on cost sharing, costs shall be shared equally.
- All sources of data are acceptable, published papers, internal reports, read-across from other related compounds, QSAR data, etc. Providing that it can be scientifically justified.
- Existing data does not have to have been generated subject to GLP and/or QA checks for it to be acceptable.
- New data generated for registration purposes shall be generated according to OECD guidelines and subject to GLP.
- If animal testing is proposed, approval for the studies must be obtained from the ECA before the work is carried out. This is to cover the eventuality that some confidential information is known to the ECA that provides information unknown to the SIEF and that could render the test programme proposed, redundant. There may be non-animal *in vitro* or *in silico* methods that would generate information acceptable to the ECA that had not been considered by the consortium involved in registration.
- Costs of any new data generated shall be shared equally by all members of the consortium.
- The participants in a SIEF shall agree the classification of the substance.
- It is anticipated that the major producer in a SIEF shall act as the lead organization although this is not obligatory.

As can be seen, there will be a lot of intra-industry co-operation and communication in relation to each SIEF and it will be a time consuming exercise on the part of those organizations or companies that take a lead role in the registration process. Time will tell whether the practicalities of co-operation match the theory as expressed in the REACH regulation. It is not a new thing however for the major EU chemicals producers to act in response to legislation in this way. What will be interesting is to see how major importers, such as retailers, react to their obligations under REACH.

3.5 Exemptions: Cosmetics?

There are exemptions from the requirements to register under REACH. These are in general related to product groups that are already subject to

their own strict regulatory requirements however, such as food, pharmaceuticals, agrochemicals, radioactive substances, biocides and to a certain
extent cosmetics.

Polymers are not subject to REACH largely because such molecules are
generally inert, however if there is free monomer present at 2% or more
then the monomer is subject to registration. Also, although polymers may
be exempt from registration, any substances added to the polymer (e.g.
plasticizers) are subject to registration; it is the polymer, not the plastic nor
the product that is exempt.

There is also a limit of one tonne per manufacturer or importer per year
for registration purposes. If the amount made or imported is less than one
tonne of each individual substance per year, then there is no obligation
to register: if there are substances of very high concern, there could be a
requirement to notify as in the pre-registration notification process, but
that is all.

Cosmetics are exempt from REACH as far *as the human health implications are concerned*. Cosmetic products imported into the EU may
contain ingredients that are not subject to the human health risk assessments required under REACH because these are already addressed in the
Cosmetics Directive,[2] which requires that each product is assessed for
safety by means of a risk assessment by a qualified individual before the
cosmetic can be placed on the market.

The environmental effects of cosmetic products are not subject to this professional assessment however and these aspects are subject to REACH.

Any importer of cosmetic products that contains substances that total in
excess of one tonne per year will be required to register those chemicals and
to generate an environmental risk assessment. This is of particular relevance
to retailers who have their own branded shampoos and hair care products
manufactured outside the EU. Many of these products utilize similar basic
ingredients and the volume of surfactants such as sodium lauryl ether sulfate, cationic conditioning agents, and so on may well exceed the one tonne
limit, triggering the REACH registration process and risk assessment.

In addition, the packaging of the cosmetics themselves, although considered articles, may contain substances that could be subject to REACH. It
will be the responsibility of each importer to determine their obligation

under REACH in respect to the goods that they import. In order to determine what obligations there are an importer will need to know:

- The total tonnage of each product that they import per year.
- The formulation of each product.
- The weight of packaging.
- The composition of packaging.

Once this information has been obtained from the suppliers, it will be the responsibility of the importer to calculate the weight of each chemical that they import with each product/package. They will then need to add up the total amounts of each chemical that they have imported over the year and if any single chemical substance totals one tonne per year or more then that substance will be subject to notification and registration. Suppliers to the EU will need to be aware that these data are going to be requested by their customers and the reason why it will not be enough to simply state that there are no substances that exceed the one tonne limit; each importer may well have several sources of the same chemical in a year and it is the total annual amount from all sources that will be important, not the amount involved with just one product.

3.6 Registration

In order to register a chemical—and remember REACH applies *only* to single chemical substances, no matter how they enter the EU—a registration dossier must be submitted to the ECA. The information that is required is related to physico-chemical properties, mammalian toxicology, and environmental effects. There are sections in the REACH regulation that detail the various pieces of information that must be addressed when submitting a dossier and it is the job of the ECA to ensure that all these headings have been addressed, by checking the completeness of each registration dossier before it goes on for detailed assessment. It must be stressed that REACH is not a box ticking exercise in the sense that all tests must be completed before a registration can be approved. That was one of the failings of the old system that led to the development of REACH, if a certain test had not been done then a registration was rejected, irrespective of whether the test was relevant or not. Under REACH it is important that each phase referred to in the regulations is addressed but that may be by a scientifically valid reason why a particular piece of work is not necessary or invalid in a particular instance. This again is related to the aim of avoiding unnecessary work and costs.

It is not my intention to go through all of the headings listed in the REACH regulation that should form the basis of a dossier; that information is best gleaned from the source regulation, which can be viewed and downloaded from the EU REACH website at: http://ec.europa.eu/enterprise/reach/ index_en.htm

References

1. Regulation (EC) No. 1907/2006 of the European Parliament and of the Council of the 18 December 2006 concerning the Registration, Evaluation, Authorization and Restriction of Chemicals (REACH), establishing a European Chemicals Agency, amending Directive 1999/45/EC and repealing Council Regulation (EEC) No. 793/93 and Commission Regulation No. 1488/94 as well as Council Directive 76/769/EEC and Commission Directives 91/155/EEC, 93/67/ EEC, 93/105/EC and 2000/21/EC. *Official Journal of the European Union* 31.12.2006, L 396/1, 1–849.
2. Council Directive of 27 July 1976 on the approximation of the laws of the Member States relating to Cosmetic Products (76/768/EEC) *Official Journal of the European Union* 27.9.1976, L 262, 169. As amended.

4

REACH: An Example of the New Paradigm in Global Product Regulation

Felise Cooper and Kenneth Rivlin[1]

Global Environmental Law Group, Allen & Overy LLP, New York, NY, USA

4.1 Introduction

A sea change in environmental regulation is impacting the global cosmetics industry.

Not very long ago, California was viewed as the world's innovator in environmental regulation, with many of its programs being mimicked by other states, the Federal government, and ultimately other countries. With the formation of the European Union (EU), however, California's pre-eminence has been challenged.

Europe is now tackling a broad range of difficult environmental issues, including climate change, energy efficiency, contaminated land cleanup standards, product stewardship, restrictions on use of hazardous substances

---

[1]Felise Cooper is an Associate and Kenneth Rivlin is a partner in the Global Environmental Law Group of Allen & Overy LLP. The views in this article are their own.

C. I. Betton (ed.), Global Regulatory Issues for the Cosmetics Industry Vol. 1, 49–54
© 2007 William Andrew Inc.

in products, product testing, regulation and restrictions of genetically modified organisms, etc. Now China, Korea, Japan, Australia, the US and, yes, California, are following Europe's lead.

One particularly interesting and exciting aspect of many of the new rules in Europe is the move away from the traditional "command and control" approach, which was the hallmark of most of the significant US environmental statutes. Rather than focusing primarily on the use of fines and penalties to discourage pollution of air, soil, groundwater, or surface waters, many of the new EU product responsibility rules focus instead on the products we make, how we make them, and what happens to them when they come to the end of their useful life. The new rules are intended to discourage, reduce and sometimes ban the use of hazardous substances in certain products, and make producers assume responsibility for their products, from cradle to grave.

These new rules also make use of the market as an enforcement agent. For example, the EU's new RoHS Directive, which became effective July 1, 2006, bans the use of certain hazardous substances, such as lead or cadmium, in most electrical or electronic equipment sold in the EU above a threshold maximum concentration level. Equipment containing too much lead cannot be sold in the EU, subject to certain limited exceptions. Accordingly, producers of equipment that is not exempt are now seeking only compliant components and parts from their suppliers, and there are now fewer and fewer purchasers of lead-containing parts.

These new rules are emerging as globalization continues at an ever-increasing pace. Companies around the world that formerly focused virtually all of their attention on domestic markets have become global. Now almost every company that manufacture products may expect their products to be sold in multiple countries. Accordingly, these manufacturers, as well as their suppliers, contractors, distributors and transporters, need to be sure their product has a smooth path to all of the markets in which they are sold. Companies throughout impacted product supply chains increasingly need to expend resources to understand often conflicting rules governing packaging, labeling, advertising and testing, approval processes for new products or ingredients, and differences in enforcement approaches in different jurisdictions. Companies that must comply with the RoHS Directive in Europe now find that their products must comply with similar, but not necessarily identical, requirements in many of the other jurisdictions in which they are sold.

One significant example of this new regulatory focus on product responsibility regulation is REACH, the new EU regulation governing the *R*egistration, *E*valuation and *A*uthorisation of *Ch*emicals (REACH). In light of the perceived failings of the current European legislation on chemicals, REACH aims to establish a comprehensive regulatory framework for evaluating the impact chemicals have on the environment and human health, and for assessing whether the most potentially hazardous of those chemicals should be subject to use authorizations or possible ban. The new regime is underpinned by the principle that it is the responsibility of manufacturers, importers, and downstream users to ensure that the substances (and products) they manufacture or place on the market are adequately controlled. While the direct impacts of REACH on the cosmetics industry are limited, REACH is of increasing interest to cosmetics manufacturers and distributors because of effects it may have on regulation that more directly impacts the cosmetics industry, and because REACH may in fact result in changes in the availability of certain substances which may be used in cosmetics now or in the future.[2]

4.2 REACH: "No Data, No Market"

> . . .substances on their own, in preparations or in articles shall not be manufactured within the Community or placed on the market unless they have been registered in accordance with the relevant provisions of this Title where this is required.

> REACH Article 5

4.2.1 What is REACH?

REACH is the new EU regulation governing the *R*egistration, *E*valuation and *A*uthorisation of *Ch*emicals. It puts the onus on industry to analyze the substance composition of virtually all products manufactured or used, and to work with chemicals suppliers and regulatory authorities to make sure that every substance (not just new substances) are properly registered

---

[2]There are of course a range of EU laws and regulations governing the cosmetics industry, including: the Cosmetics Directive (76/768/EEC), Directive 95/17/EC on ingredient labeling, Dangerous Substances Directive (76/769/EEC), Regulation (EC) No 2037/2000 on ozone-depleting substances in aerosol products, Packaging and Packaging Waste Directive (94/62/EC), Prepacked Products Directive (80/232/EEC) (revision proposed) and the Aerosol Dispensers Directive (75/324/EEC). These are topics for another day.

and, as applicable, evaluated, authorized, and/or restricted, before any substances or finished products are put on the market in the EU.

The central provision, quoted above, prohibits placing on the market both chemicals and finished goods unless those products comply with REACH.

REACH has been officially passed and will enter into force on June 1, 2007, though most of the operative provisions will be phased in over a series of years.

REACH transfers the cost of data collection and pre-market assessment of chemical substances to industry—previously, such analysis and assessment had been carried out by Member State authorities based on information provided by industry. After extensive study and lobbying, European legislators and authorities determined that the existing system was inadequate and that there was a need for a harmonized approach—hence REACH was born. REACH is based on the "precautionary principle" that no action should be taken until it is proven that harm will not result and aims to establish a comprehensive framework for analyzing the impact chemicals may have on the environment.

4.2.2 How Does REACH Affect the Cosmetics Industry?

REACH requirements mainly target companies that manufacture chemicals and certain REACH provisions do not apply to the use of substances in cosmetics, but REACH does contain provisions directed towards producers and importers of *substances*, *preparations*, and *articles* as well as *downstream users* of substances and preparations.

In addition, it is expected that REACH will limit the range of chemicals and products currently available in the EU. The costs of certain chemicals are also likely to increase. Although REACH obligations will be phased in over a series of years, companies are already assessing the extent to which the chemicals they manufacture, import or use are caught by REACH. Awareness of the impacts REACH will have on your business, together with professional support at an early stage, will be critical in helping you to successfully manage the challenges and opportunities arising from REACH.

4.2.3 How will REACH Work?

Under REACH, manufacturers and importers of a vast range of substances made or brought into the EU in quantities of over one metric ton per year will have to register their substances with the new European Chemicals Agency. A failure to register will mean that substances cannot lawfully be placed on the EU market. It has been estimated that to register a chemical could cost in excess of €1 million. However, to keep costs (as well as animal testing) to a minimum, companies needing to register chemicals will be allowed to work together in consortia designed to maximize the benefit of data sharing between participants and preserve a "one substance, one registration" principle.

There will be a phased approach to the registration of certain substances. Chemicals imported or produced in quantities of 1,000 metric tons per year or more, as well as other high-risk substances, will be registered first. The next wave of registrations will be broadened in scope to capture quantities of between 100 and 1,000 metric tons per year, and so on. To take advantage of the phase in, companies will need to pre-register substances starting in June 2008.

Applications for registration of a substance will have to include a wide range of technical information, including details of the hazards arising from the substance, proposals for further testing and guidance on the safe use of the substance.

A number of chemicals will need to be expressly authorized under REACH in order to be used in the EU. This has led to concerns among companies as to whether substances they are currently using or manufacturing will, for the first time, be phased out.

4.3 What Should Companies be Doing Now?

- Identify who is responsible internally for managing the impacts from REACH, including which parts of the business need to be involved, and develop an implementation plan.
- Prepare an inventory of substances that may be affected by REACH (e.g., substances in your products for sale for the EU or used in your operations in the EU).
- Engage in dialog with your suppliers and customers as to how they expect REACH to be managed.
- Ensure that your chemical suppliers are including your use in their registrations.
- Prepare for the potential that some substances you currently use could become unavailable or more scarce due to increased scrutiny, possibly requiring product design changes.

- Assess whether you may need to appoint an EU representative.
- Review and update your contracts with suppliers and customers to address REACH responsibilities.
- Assess the financial impact of REACH compliance on your business and examine whether any disclosures may be needed in public reports.
- Evaluate the potential antitrust implications of entering into discussions with competitors and suppliers.

References

The REACH Regulation (Number 1907/2006) published in the Official Journal: http://eur-lex.europa.eu/LexUriServ/site/en/oj/2006/l_396/l_39620061230 en00010849.pdf

European Chemicals Bureau's web page on REACH: http://ecb.jrc.it/REACH/

European Commission's web page on REACH: http://ec.europa.eu/environment/chemicals/reach/reach_intro.htm

5

Developing a Global Regulatory Strategy: Leveraging Local Knowledge to Drive Rapid Market Entry

Neil L. Wilcox

Global Regulatory and Scientific Affairs, Kimberly-Clark Corporation, Neenah, WI, USA

5.1 Introduction

The cosmetic industry faces multiple challenges when attempting to market its products around the world. International companies are many and varied, and all face regulatory obstacles. Although multiple strategies may be employed, no single model is necessarily considered the best. Multiple variables must be taken into consideration prior to developing a strategy that is the most favorable for a particular organization. In fact, "leveraging local knowledge" is an approach that, by some accounts, may seem contrary to the trend for multinational companies to globalize as they reorganize into a flatter, more flexible matrix model. To some, globalization imparts the notion that with appropriate information technology, the international company should be able to reduce its physical presence around the world. Of course to a large extent, this is true as many embrace the approach of doing "less-with-less." In other words, developing a strategy that leads to a

C. I. Betton (ed.), Global Regulatory Issues for the Cosmetics Industry Vol. 1, 55–62
© 2007 William Andrew Inc.

more highly trained workforce that is competent in using new technology and that has the net effect of reducing overhead and increasing capacity. For many valid and important reasons, the successful international cosmetic company must be familiar with multiple facets of the external environment to successfully navigate various contributing factors to develop a successful regulatory compliance program in any particular geographic region.

5.2 Globalization: "One Size Does Not Fit All!"

Any international company wishing to expand its portfolio of products and markets endeavors to reduce the variables in its products resulting in that ever elusive "global product." Of course, such an approach reduces costs and is an important part of any global marketing strategy. Focusing on this obvious and somewhat intuitive approach leads to inefficiencies and missed opportunities. Although products intended to be marketed globally should be developed to the extent possible to meet requirements in multiple countries, meeting this criterion is for all intents and purposes impossible. Differences in labeling, packaging, and allowable ingredients obviate the ability to market the elusive ubiquitous product. The intelligent approach is to minimize differences, which clearly reduces costs and maximizes profits. In many corporate cultures of all types and sizes, product safety, regulatory affairs, corporate quality, and clinical operations are considered "cost centers," which lead business profit centers to limit these resources in favor of the "revenue centers." This common but naïve approach encourages those responsible for product development and marketing to limit their interaction with the teams responsible for ensuring product safety and compliance. In other words, when products are being developed and claims are under consideration, those responsible for ensuring product safety and complying with local market laws are often not consulted early or often enough in the product life cycle for several reasons.

5.3 Using Regulatory Compliance for Competitive Advantage

A product's regulatory classification provides the basis for determining market requirements around the world. This simple fact may be used to not only guide the development of the product and its intended benefits, but also establish the go-to-market path that drives innovation and increases speed-to-market. Commonly and predictably, those responsible for product development focus on innovating existing company products, creating products that are new to the company and those that may be new to the world. Included in this paradigm are those responsible for marketing who are trying to establish an overall presentation of the product that appeals to the customer, shopper, chooser, and user. There are multiple, complex and sophisticated variables that impact these decisions and provide the basis for products to either succeed or not meet expectations. Undeniably one of the realities successful cosmetic companies face is that their products must constantly change to achieve corporate growth. The pressures that drive profit tend to obviate involvement of the product compliance and safety functions in early product development.

All those in the business understand the "Dr. No" phenomenon, which in part contributes to the reluctance of product developers to engage compliance functions early in product development. Developers and marketers need to constantly push the envelope to create unique and successful products. As a result, there is the tendency to create the formulae, claims, artwork/labeling and packaging, and as a final step check with product safety and regulatory to make sure all requirements have been met. The missed opportunities from this approach are many. The globalization phenomenon is moving fast and requires international companies to be more "matrixed" (viz., less hierarchical), which include developing processes and systems that drive constant innovation and increased speed-to-market. One of the hidden and often unappreciated advantages available to businesses is expertise in regulatory affairs, product safety, corporate quality, and clinical affairs. (Clinical affairs in this context refers to designing and managing studies conducted in human beings for the purpose of meeting regulatory requirements.) Properly engaged, these functions which heretofore have been treated as necessary "cost centers" can be turned into "revenue centers." The *sine qua non* to developing a global compliance infrastructure that drives innovation and increases speed-to-market is made up of the following dimensions: (1) organizational design, (2) targeted capability, (3) local knowledge, (4) business engagement, (5) crossfunctional processes, systems, and tools, and (6) communication plan.

5.3.1 Organizational Design

Establishing an organization structure that meets the needs of a global company is dependent upon many variables including the type of products sold, the countries in which they are marketed, the regulatory classification of the products, and the local regulatory requirements. It is imperative to maintain organizational flexibility and establish the ability to meet changing product types and optimize speed-to-market requirements. One constant in the current business milieu is that of change. Market pressures require product innovation and penetration change on a regular basis; the successful global organization footprint must be nimble and move with the business plan. An efficient approach to globalization does not mean centralizing an organization geographically. The successful centralized organization needs to be strategically placed and act centralized.

5.3.2 Targeted Capability

It is exceedingly important that the appropriate talent, capability, and expertise are in place to support the business needs. Talent management is a fundamental business requirement and is no less important in establishing the infrastructure of a global compliance model. Identifying the appropriate balance between experience and formal education requires special attention. It is important to note that these resources are not necessarily internal to the business but may be outsourced as well. Fundamental to building capability is a strong talent management program that maintains a high priority for recruiting relevant expertise, continuous training for existing staff, and ensuring a strong career progression program. Most importantly, capability building must target current and future business needs based upon planned portfolio growth and market penetration. Clearly, this requires a flexible and targeted strategic approach to building capability.

5.3.3 Local Knowledge

The fundamental precept to creating a successful global regulatory compliance model is that of understanding the local market conditions. This includes, but is not limited to, knowing the product regulatory requirements, understanding how the regulatory authorities implement these requirements in the local markets, being sensitive to local customs and societal differences, and establishing adequate resources for language translation. This approach is not necessarily a return to the decentralized model perceived as adding layers and personnel. Instead, it is an intelligent and thoughtful approach to leveraging local knowledge to drive global innovation and reach. As mentioned earlier, the notion of "globalization" does not mean using technology to reduce global presence. It means using technology to drive efficiencies in improving communication and reducing costs. This may be done through outsourcing and offshoring, but most important is knowing the market characteristics, including multiple dimensions of regulatory intelligence.

5.3.4 Business Engagement

One of the most significant weaknesses in regulatory organizations is that they do not adequately understand their own businesses. Their approach too often focuses on knowing the regulations. Although this may intuitively make sense, it does nothing more than create a reactive organization that is perceived as "Dr. No" because it has the tendency to focus on "informing" the business of requirements that need to be met. Unfortunately, this creates not only an adversarial relationship between the product development process and those responsible for ensuring regulatory compliance, but also clearly establishes "regulatory affairs" as a "cost center." Properly implemented, the intelligent, best practice, global regulatory organizational model engages with the business early in product development and becomes a proactive business asset, which will then translate into a "revenue center." To accomplish this, the regulatory team must have a comprehensive and substantial understanding of the business. Only through seeking to understand the insights, research, customer needs, and intended benefits that the business is trying to capture as part of product innovation can the regulatory experts be perceived as value added to the business.

5.3.5 Cross-Functional Processes, Systems, and Tools

One of the clear advantages of the global regulatory infrastructure is providing efficient and effective standardized processes, systems, and tools. The global regulatory model needs to be a work stream to the international company. In other words, it needs to work as a cross-functional entity providing a comprehensive service across many dimensions of the business. To suggest common examples, research and development, manufacturing, marketing, artwork/labeling, legal, and supply chain all need to communicate with the regulatory organization in an efficient, standardized fashion so as to contribute to product innovation and increase speed-to-market.

5.3.6 Communication Plan

Never to be taken for granted is the need for effective communication, which really means developing leadership skills at all levels in the organization. There is nothing novel about this concept; however, effective communication is one of the most difficult challenges in any corporation, and this is especially true in the regulatory affairs arena. Regulatory personnel tend to have scientific and technical backgrounds, which increases the likelihood for them to focus on regulatory requirements and what the business needs to do to meet them. The more effective paradigm is to seek to understand how the business works, the intended benefits of their product concepts and proactively contribute to product innovation. Paramount to successful communication is the ability to listen and ensure that there is a thorough understanding of the business' strategic business portfolio planning. Only by understanding the complexities of fulfilling insights and the underpinnings of product development can the regulatory organization transform itself into a value-added and coveted entity. It must be well understood that without an effective communication plan, the best organizational design, targeted capabilities, local knowledge, business engagement, and processes, systems, and tools, an effective global compliance program will not be established. In this context, leadership means setting direction, aligning constituencies, motivating and inspiring, and as a result producing change. The effective communication plan, therefore, will motivate the business partner to work proactively with the regulatory team because it is well understood that it will contribute to meeting the business objectives.

5.4 The Global Launch: "Driving with Insights and Regulations"

Many companies selling products in multiple countries are by definition international companies. But such a company is not necessarily a "global" enterprise. Similarly, launching a product sequentially around the world without adequate partnering with regulatory compliance planning early in product development will cause a collision with onerous regulatory requirements. This approach will predictably obstruct market entry and reduce profits. Globalization means taking advantage of economies of scale provided through work streams as found, for instance, through the global regulatory model that leverages local knowledge. To establish such an organization as a proactive advantage to meet the business needs, it must create an infrastructure that supports global launches. A symptom of the international company that is not "globalized" is one which develops a product in an isolated geographical location and then rolls the product to foreign markets in serial fashion. This is often done without adequate knowledge of the local product consumer preferences or advanced knowledge of the market's regulatory requirements and peculiarities to market entry. As any company that has attempted to create the perfect "global product" has learned, there is no such thing as a product, including its claims, formula/materials, packaging, and labeling, that can be sold in all markets worldwide. In particular, the more regulated a product becomes, the more this phenomenon becomes an undeniable obstacle to worldwide distribution.

An effective global launch model starts with developing an approach that results in understanding the differences in market preferences through insights and market research. Once the characteristics that must be met for a product to be successful in multiple markets are understood, the truly global product may be developed. Similarly, without understanding regulatory and market entry requirements prior to product development, a successful global product cannot be made and global launch cannot be achieved.

Early in product development, the potential benefits of the product (potential claims) must be identified along with the intended markets. Clearly, a global launch cannot be planned without knowing in advance the markets in which they may be sold. Similarly, without understanding the intended claims, the regulatory requirements cannot be confirmed. The undeniable conclusion is that the successful global launch model is dependent upon the regulatory team working closely with product development, marketing, legal, supply chain, etc., so that there may be a thorough understanding of the intended claims and markets. Once these key elements are known, the global regulatory organization, strategically aligned with the business and geographical markets, may go forward with conducting regulatory assessments to determine the requirements for market entry. Once these requirements are known, based upon the regulatory classification for each intended market, strategic business decisions may be made that result in not only knowing the markets in which the products will be sold, but also required timelines founded upon the actions necessary to meet regulatory requirements. Of course, once the decision has been made as to the marketing strategy based upon product claims, then the entry requirements are known for each global market. The launch date for each market may then be determined, and the time necessary to complete the multi-dimensions of regulatory mandates. These will include product testing for safety, clinical testing, compiling premarket dossiers, claims substantiation, meeting manufacturing quality requirements, and finalizing artwork/labeling. A global launch does not necessarily mean the product will be simultaneously launched in all markets. However, it would be quite feasible for the product to be launched at the same time in multiple markets and then rolled out in a preplanned sequence based in large part on the time needed to complete regulatory registration and other requirements. The global strategic launch plan leverages "local knowledge to drive rapid market entry," illustrating a "best-business" approach.

5.5 Conclusions

The business that develops global products which incorporate market preferences and claims that are winners in the selected markets will have the competitive advantage. Moreover, the international company that determines regulatory classification for each intended market and develops a global strategy early in product development based, in part, on the timelines and resources required to meet regulatory requirements, will employ the advantages of "leveraging local knowledge" as a global strategy. The favorable result of this paradigm will be to establish the product regulatory and safety compliance program as a revenue center and an asset that contributes to both top and bottom line growth.

6

Cosmetics: Toxicity and Regulatory Requirements in the US

Harold I. Zeliger

Zeliger Chemical, Environmental and Toxicological Services, West Charlton, NY, USA

6.1 Introduction

Before addressing regulatory requirements and warnings for cosmetic products, we need to address the issue of cosmetic toxicity. Products that are not toxic do not require hazard warnings. Many cosmetics, however, contain components that are toxic to skin, eyes, when inhaled or when ingested.

Cosmetics are formulated with inorganic and organic chemicals, many of which are toxic. Table 6.1 shows a partial list of toxic chemicals commonly incorporated into cosmetics sorted by function.

The chemicals listed in Table 6.1 are toxic to the skin, eyes, and respiratory tract. Some, propylene glycol for example, have been identified as sensitizers following long-term exposure. Formaldehyde and butylated-hydroxyanisol (BHA) are considered to be carcinogenic.

C. I. Betton (ed.), Global Regulatory Issues for the Cosmetics Industry Vol. 1, 63–70
© 2007 William Andrew Inc.

Table 6.1 Partial List of Toxic Chemicals in Cosmetics

Preservatives	**Antioxidants**
Formaldehyde	Butylatedhydroxyanisol
Quaternium 15	Butylatedhydroxytoluene
Methyl and propyl paraben	
Ointment bases	**Emulsifying Agents**
Lanolin alcohol	Sodium lauryl sulfate
Propylene glycol	Glyceryl stearate
Polyethylene glycol	Cetyl alcohol

In addition to these, pigments comprising inorganic minerals and organic binders, organic dyes (e.g., azo dyes), and fragrances are incorporated into cosmetics. Fragrances have recently come under scrutiny in European Union (EU) regulations. This will be discussed later.

Hair treatment products contain several sensitizing compounds (including ammonium thioglycolate) and corrosives (such as sodium hydroxide).

Mixtures of toxic chemicals present dangers not always predicted by the toxicities of the individual compounds. My research has shown that when people are exposed to mixtures of lipophilic (oil soluble) and hydrophilic (water soluble) compounds, the lipophiles facilitate the absorption of the hydrophiles across the body's lipophilic membranes and produce unanticipated effects. These include:

1. Enhanced toxicity: Toxicity beyond that predicted by the additive effects of the individual species.
2. Low-level toxicity: Toxic effects below the no effect observed levels (NOEL) of any or the individual components.
3. New target organ attack: Attack on organs not known to be impacted by the individual compounds.

Examples of each of these are as follows:

1. Enhanced toxicity: Painters applying latex paint developed reactive airways dysfunction syndrome (RADS), a condition not predicted from the known toxicities of the chemicals in the paints.
2. Low-level toxicity: Shoemakers exposed to a low-level mixture of *n*-hexane, cyclohexane, MEK, and ethyl acetate

experienced neurotoxic effects. These were not predicted at the concentrations of the chemicals to which the exposure took place.

3. New target organs: The offspring of both male and female workers in an electronics manufacturing plant who worked with a large number of lipophilic and hydrophilic chemicals had a cluster of brain cancers. Brain cancer is not known to be caused by any of the chemicals involved.

Do chemical mixtures in cosmetics give rise to unexpected effects? Most certainly. Cosmetic products are composed of many lipophilic and hydrophilic chemicals, and one should not be surprised when users of these products report "strange" afflictions following their application. I have investigated numerous instances of injuries from the use of cosmetics and personal care products. In many of these cases, the injuries sustained could not be easily accounted for by a consideration of the individual chemicals involved.

As stated above, the toxic effects of chemical mixtures are often not predictable. Mixtures often behave as a new species, completely different from the individual components. This points out the need to test all new formulations prior to their introduction to the marketplace. This also points out the need to retest whenever any one component is replaced with a substitute, because new mixtures are created. There are numerous examples of reformulated products having greater toxicity than the originals. Ones I have examined include moisturizing creams, shampoos, hair sprays, and salon hair products.

6.2 Regulatory Requirements for Cosmetics

Cosmetics marketed in the US, whether manufactured in the US or imported must comply with labeling requirements of the Federal Food, Drug and Cosmetic (FD&C) Act, the Fair Packaging and Labeling (FP&L) Act, and the regulations published by the US Food and Drug Administration (FDA) under the authority of these laws.

FDA regulations [21 CFR 701.2] provide labeling compliance information. FDA provides a cosmetics labeling manual on the web (www.cfsan. fda.gov/~dms/cos-lab3.hmtl). I am certain that all of you are familiar with the labeling regulations. What is of interest here are the warning statement and ingredient declaration sections.

6.2.1 Warning Statement

I quote from the FDA:

> The safety of a cosmetic may be considered adequately substantiated if experts qualified by scientific training and experience can reasonably conclude from the available toxicological and other test data, chemical composition, and other pertinent information that the product is not injurious to consumers under conditions of customary use and reasonably foreseeable conditions of misuse.

FDA goes on to say,

> The safety of a cosmetic product can adequately be substantiated by:
>
> a. Reliance on available toxicological test data on its ingredients and on similar products, and
> b. Performance of additional toxicological and other testing appropriate in the light of the existing data.

FDA concludes, "Even if the safety of each ingredient has been substantiated, there usually still is at least some toxicological testing needed with the formulated product to assure adequate safety substantiation."

What is not defined is which tests are to be carried out to establish safety. This is in sharp contrast to the requirements of the Federal Hazardous Substances Act (FHSA), which is enforced by the Consumer Product Safety Commission (CPSC) and the Occupational Safety and Health Administration's (OSHA's) Hazard Communication Standard. Those regulations, as well as the Environmental Protection Agency's (EPA's) Federal Insecticide, Fungicide and Rodenticide Act (FIFRA) detail precisely which toxicological testing is required.

FHSA and OSHA require that the identity and toxicological properties of any chemical present in 1% or more (or 0.1% if it is a carcinogen) be listed. Cosmetic products, therefore, have laxer reporting requirements than consumer or industrial chemicals. FDA regulations are particularly lacking when it comes to identifying flammability of cosmetic or personal care products. Again FHSA and OSHA have precise definitions and requirements.

6.2.2 Ingredient Declaration

FDA requires a listing of ingredients in descending order of predominance. There are two notable exceptions:

1. Fragrance and flavor ingredients need not be identified other than as "fragrance" and "flavor."
2. Trade secrets need not be divulged. FDA defines a trade secret as follows: "A trade secret may consist of any formula, . . ., which is used in one's business and which gives him an opportunity to obtain an advantage over competitors who do not know or use it."

CPSC, OSHA, and FIFRA do not allow such exemptions. California's Proposition 65, which will be discussed shortly, requires that chemicals that are carcinogenic or induce birth defects be listed on the label.

I will not address the legal issues associated with trade secrets, other than to say that it allows for the withholding of product formulation information. Such withholding permits the true identity of potentially toxic chemicals to be concealed.

6.3 California Proposition 65 and California Safe Cosmetics Act of 2005

California Proposition 65 requires the publication of a list of chemicals known to the state to cause cancer or reproductive toxicity. It reads as follows:

> The Safe Drinking Water and Toxic Enforcement Act of 1986 requires that the Governor revise and republish at least once per year the list of chemicals known to the State to cause cancer or reproductive toxicity.

There are currently about 750 chemicals on the Proposition 65 list. Two of the chemicals in Table 6.1, formaldehyde and butylated hydroxyanisole (both carcinogens), are listed in the Proposition 65 list.

The California Safe Cosmetics Act of 2005 requires the identification of toxic chemicals in cosmetics.

6.4 EU Cosmetics Regulation

Cosmetics in the EU are regulated primarily through the European Cosmetics Directive and its Seventh Amendment, which went into effect in March 2005. It is currently in force all over Europe. The Seventh Amendment bans the continued use of three classes of toxic chemicals. These are as follows:

1. Those which pose risks of cancer.
2. Those which cause endocrine (hormonal) or reproductive disturbances.
3. Those which cause genetic damage.

The Seventh Amendment also requires the labeling of fragrance ingredients which can cause allergic reactions, contact dermatitis and asthma in sensitized users. Currently, 26 fragrance ingredients require such labeling. These are listed in Table 6.2.

6.5 The Future of Cosmetics Regulatory Requirements

The future of cosmetics regulatory requirements is now upon us. The net effect of California Proposition 65 and the EU Seventh Amendment will put great pressure on cosmetics manufacturers world wide to conform with these. In the US, FDA may be slow to require compliance with Proposition 65

Table 6.2 Fragrance Ingredients Requiring Labeling Under the EU Seventh Amendment

Amyl cinnamal	Benzyl alcohol
Cinnamyl alcohol	Citral
Eugenol	Hydroxycitonellal
Iso-eugenol	Amylcinnamyl alcohol
Benzyl salicylate	Cinnamal
Coumarin	Geraniol
Hydroxyisohexyl-3-cyclohexene	Carboxyalhehyde
Anise alcohol	Benzyl cinnamate
Farnesol	Butyl methylpropianal
Linalool	Benzyl benzoate
Citronellol	Hexyl cinnemal
D-limonene	Methyl-2-octynoate
Alpha-isomethyl Ionone	Evernia prunastri extract

and the Seventh Amendment, but the marketplace and our litigiousness may greatly accelerate the process.

In the interim, cosmetics manufacturers can take the time still available to review and evaluate the toxicities of all their formulations and replace toxic components where indicated. I recommend that all the revised formulas be tested thoroughly for toxicological properties following the guidelines established by FHSA and OSHA. Manufacturers are already familiar with OSHA regulations from the requirement to produce Material Safety Data Sheets (MSDS) for all their products for the protection of their employees and others who come in contact with these products in the course of shipping, warehousing, and selling them.

Having said what I just did, what are the choices available to the cosmetics industry?

1. The government (i.e., FDA) will ultimately be forced to extend their regulations. But, FDA does not want to regulate. FDA is poorly equipped to oversee new regulations. I quote from an article in the August 14, 2006, issue of *Chemical and Engineering News*: "FDA is a regulatory agency that does not want to accept the fact that it is a regulatory agency." This quote was made in the context of the toxicity of acrylamide in foods, but applies, in my opinion, to cosmetics as well.
2. FDA regulates selectively. Just recently, FDA proposed a ban on skin bleaching creams containing hydroquinone, citing it as a possible carcinogen. It has been known since 1988 that hydroquinone is percutaneously absorbed. It also well established that hydroquinone is a precursor to benzoquinone in the body and that benzoquinone is the metabolite of benzene known to be a leukemogen.

So why did FDA wait so long to act against hydroquinone?

FDA like CPSC, EPA, and other regulatory bodies are not equipped to investigate all potential situations. They react to events and political pressure. The problem is, one never knows when they will intervene.

What can the cosmetics industry do?

Be proactive rather than reactive. Do it yourself. Set your own standards through CTFA. Present FDA with a new set of testing and labeling standards and work with FDA to establish your standards.

How?

Set up an industry committee, staff it with professionals and have it draft recommendations for industry consideration. You can do this on a company by company basis, but I do not believe such an approach will be as effective since you will be pitted against each other. I believe FDA will be open to responsible recommendations. Perhaps FDA can be persuaded to formally review product labels as EPA does for pesticides under FIFRA. As a result of EPA's role, there have essentially been no successful law suits against pesticide manufacturers in many years.

There are other precedents for industry establishing standards. The Chemical Manufacturers Association (CMA) has adopted a standard called responsible care that covers all aspects of chemical manufacturing, distribution, use and discarding of chemicals. It basically assures the safe use of chemicals from cradle to grave. Almost all chemical companies subscribe to it and it has elevated and unified the treatment of chemicals. The paint and coatings industry has also done something similar in the labeling context.

The other alternative, do nothing and let the EU, Proposition 65 and perhaps the FDA set the new standards. Along with this, litigation will surely drive the issue. Litigation is driving pharmaceutical testing, labeling, and marketing. Successful litigation against pharmaceuticals is decaying confidence in FDA. Along with the EU's regulations, Proposition 65 and this anti-FDA psychology, inaction on the part of the cosmetics industry will most certainly encourage more litigation against the cosmetic industry.

I know, however, from my experience that if the industry establishes, and its members practice, the state of the art, there will be fewer injuries to users of its products and fewer successful law suits even if the government is not involved. CMA's responsible care is a good example of this.

Finally, it should be noted that the state of the art of toxicological knowledge and, therefore, labeling requirements is a moving target. Ongoing research is constantly expanding our knowledge of toxicology and necessitating an ongoing review of warning label content.

In conclusion, the future will certainly bring more restrictive testing and labeling regulations for cosmetic products. By being proactive rather than reactive, I believe the industry can drive the regulatory process rather than having to defend against it.

7

Restricted Substances in Consumer Products: The Challenge of Global Chemical Compliance

Rudolf A. Overbeek and Joel Pekay

Intertek Group, Houston, TX, USA

7.1 Introduction

The evolution and expansion of trade has become a driving force in our growing world economy. World trade creates easier global access to developing technologies and new products. The increasing demand for technological innovation paired with the growing trend for fashionable gadgets is fostering new environmental regulations and increased worldwide concern for all entities from humans, animals, to the entire ecosystem regardless of global region.

Governments, politicians, non-governmental organizations, consumers, and businesses worldwide are therefore demanding that substances, either known to be (potentially) hazardous or expected to be harmful, will become regulated or restricted in everyday consumer products. Because new legislations are becoming more prominent and widespread and since

C. I. Betton (ed.), Global Regulatory Issues for the Cosmetics Industry Vol. 1, 71–81
© 2007 William Andrew Inc.

environmental compliance has been enforced strongly, it is expected that in the years to come new environmentally driven chemical regulations will substantially affect the consumer products market and that a wide range of substances will be added to an ever-growing list of restricted or regulated substances. On top of that, there are regional discrepancies in the severity of environmental regulations imposed and industry reaction to these regulations. Considering all the regulations imposed or awaiting legislative approval across the globe, "regulation overload" becomes apparent.

As an example, under REACH (Registration Evaluation and Authorization of CHemicals), to be enforced in the EU from June 1, 2007, manufacturers, retailers, brand managers, traders, and distributors are required to demonstrate that:

1. The use of a chemical substance does not adversely affect either human health and/or the environment by providing documented safety information on the substance in question.
2. All risks for potentially causing injury and/or health concerns to both humans and the environment by the chemical itself are identified and managed.

REACH has far-stretching implications, since it covers:

1. Substances (i.e., single chemicals – e.g., a *Solvent*)
2. Preparations (i.e., mixtures—consumer products and industrial formulations – e.g., *Inks* to be used in pens or cartridges)
3. Articles (i.e., anything where the shape or surface properties is of greater importance to the function than chemical composition itself – e.g., a *Pen* containing the ink)

Environmental supply chain management and chain of custody control will therefore become substantial, since these growing requirements will affect any chemical used, as substance, in formulations and in consumer products. The appropriate alignment of Restricted Substances Policies and Environmental Compliance Policies of each brand and producer with their supply chain and their supply chain business processes is therefore crucial.

In consideration of the above defined challenges and based on Intertek's experience with the supply chain, the subject article summarizes the overall approach Intertek recommends.

7.2 Addressing the Challenges of Emerging Globally Restricted Substances Regulations

Smaller firms will suffer more from the higher costs required for compliance to current and future regulations. But that is not to say that large corporations will go unscathed. On the contrary; due to the high recognition, reputation, and respect of the brands owned by these corporations, one slight mistake in implementing the required regulations or publicly rejecting the "reasonable steps" asked of your company can tarnish your brand forever. The trust, loyalty, and expectations of the customer center on a corporation's ability to leverage their brand against all odds in favor of its target market.

But how do corporations master such an incredible feat? More and more corporations are relying on the expertise and reputation of outside companies to help them sort through the "confusion of compliance". A sound partner must be a "one-stop-shop" providing vendor training, global laboratory coverage for materials and components certification, environmental and quality compliance audits, and a 24/7 library of materials, components, products and vendor compliance data. In addition, such a company must be able to provide the *3 Cs to compliance*—cost control, current information on all regulations proposed in every region, and a confidential relationship between client and company.

Intertek works with advisory teams and local experts to develop innovative solutions for renowned corporations across the globe. Intertek supports major brands in developing successful business approaches defining their expectations with complete clarity to suppliers and associates alike, through educational programs, effective monitoring systems, and continuous improvement processes.

A strong strategic partnership with a global network of supply chain control and monitoring specialists will ensure an ongoing success of the quality, compliance, and safety processes, which are key components of the overall business process enabling sustainable growth, currently and in the future. Key ongoing benefits should include:

- Connectivity to a truly global network dedicated to product quality, compliance management, safety, and innovation.
- Access to state-of-the-art regulatory research tools that can assess and help predict the need for necessary actions.

- Access to collective research data and emerging knowledge of the world's leading experts in product compliance, compliance management, safety, and innovation.
- Ability to deal with compressed timetables and elimination of production setbacks, rework, product failures, recalls, and returns.

It is important to develop a solution that will cost effectively allow the *proactive* management of restricted substances throughout product lines taking into account changing global legislation and good practices/voluntary standards as well as flow of information within the supply chain. As a seller or manufacturer of a consumer product it is therefore highly important to have an appropriate visibility of the supply chain so that it can be known which chemicals are used in the formulations, materials, and parts applied in producing the final consumer products which are sold to the consumer.

Within Intertek we have developed systems and solutions to optimize the assessment of compliance in vendor/supplier materials and information, while at the same time achieving efficiency gains for our customers. More importantly we establish consistency, risk avoidance, and cost reduction using these solutions. Through increased visibility brands and retailers will be able to perform timely assessments against legislation and customer-enforced requirements for both new products (during the design phase) and products on the market. It will allow our branded customers to transfer the responsibilities to *their* suppliers/vendors of assemblies, materials and raw materials, for entering the required compliance-related information while at the same time having the full ability to track and assess their performance and monitor compliance. This information should be complemented with brand or seller-initiated risk-based conformance verifications on vendors and suppliers, their products, assemblies, materials and raw materials, and/or on final products.

7.3 Restricted Substances: Strategy Definition

Individual businesses must determine which departments need to be involved in a restricted substances strategy. The following are examples of departmental responsibilities:

- Research and development: ensures that materials used in manufacturing contain allowable amounts of restricted substances.
- Procurement: ensures that suppliers provide base materials or products meeting all restricted substances laws.

- Manufacturing: implements procedures that eliminate introduction of non-compliant materials into the production process.
- Distribution: ensures compliant products are shipped to the correct location.
- Sales and marketing: communicates compliance to government enforcement bodies. Compliance can also be positioned as a competitive advantage when selling products to customers, retailers, or distributors.
- Executives: oversee this process and assume corporate liability.

Externally, companies should manage their suppliers that face similar decision making processes. Customers, distributors, and retailers are requiring proof of compliance from the supplier. Since standards do not exist, there is much variation in the level of documentation provided.

Some suppliers provide a self-declaration while others provide certificates of compliance. Additional testing reports may be provided although the results vary based on the test methods. There are a number of steps that can be taken to minimize business risk.

7.3.1 Steps to Meeting Global Compliance Directives

1. Prepare a strategy
2. Setup a compliance management system
3. Collect documentation and proof
4. Measure risk
5. Communicate

The first step a business should take is to create an appropriate strategy. This is achieved by creating a plan that includes the gathering of existing compliance data and identifying gaps. All departments should be involved in order to determine the impacts of the strategy, each teams needs, and how changes affect all parties involved.

As a strategy is developed, documentation is necessary as a proof of due diligence. Files should be readily available to all employees, customers, distributors, inspectors, and stakeholders. This assists with communication among employees ensuring the strategy that is established is followed.

The next step is to establish auditable compliance management systems enabling the business to track products and components, coordinate registration and test data, and manage suppliers. Compliant products should be given a new, unique product number providing inspectors, distributors, and customers an easy way to identify *compliant* merchandise. It also allows the business to ensure that materials or products received are compliant. Process management begins with the purchasing of materials or components and ends when finished products are delivered to the retail shelf. All areas in-between must be managed. Otherwise, there is a risk of introducing a non-compliant product.

Thirdly, the business collects all information concerning the strategy plus any material and product data to create a so-called product technical compliance file. This file should at the minimum include supplier declarations, supplier strategies, and supporting evidence such as test reports or third party certifications. The documentation is proof of the business' due diligence. If supplier declarations indicate the presence of controlled substances, the business should assess whether the regulated limits are exceeded. Concerns regarding the accuracy of a declaration should be tested and monitored for actual restricted substances levels. Finally, perform audits and random product inspections to ensure all strategies are followed.

Supplier audits are another method to ensure vendors are meeting set policies. These can be conducted through scheduled visits and on an *ad hoc* basis. A supplier that fails to comply directly impacts the business compliance and therefore their revenue and reputation.

Fourthly, measure product risk by conducting risk assessment for the presence of substances of concern based on statistics, historical information, and supplier performance. Determine which materials are most likely to contain restricted substances, identify alternative materials, and make changes to correct substance levels. Based on the results, the business will learn how much risk it needs to mitigate, the potential for lost brand value, lost revenues, and increased insurance costs due to non-compliance.

Finally, communicate the strategy and process throughout the organization. Each employee, department, and division must know their role and responsibilities. It is vital to create continuing education programs so employees are knowledgeable about ever-changing environmental legislation. This will ultimately create a cohesive team working as one to meet global compliance requirements thereby protecting the company.

According to Intertek's experience, basic compliance documentation, submitted by the supply chain, should at the minimum include:

- Contact information: point of contact within the organization that is responsible for meeting the restricted substances requirements.

- Approach to compliance: this should be a general overview of any compliance systems that the company has in place and which are suitable for assisting compliance with any restricted substances requirements including those identified by their customers.

In addition, each supplier should be able to provide an overview of the data quality systems they have in place. This is especially relevant in those cases where the producer or brand relies significantly upon supplier information to demonstrate compliance. The systems could include risk assessments, acceptance criteria, purchasing procedures and any other relevant documentation. It may be a combination of both process-based and product/part-based documentation. This information should furthermore be supported by:

- Product-based technical documentation: typical information relating to a product's attributes that ensures compliance of a specific product.
- Producers' or suppliers' warranties/certificates: declarations that the use of restricted substances is within the permitted levels.
- Producers' or suppliers' completed materials declaration: – these declarations may be limited to the restricted substances lists identified by the respective customer.

It has also been proven highly desirable to obtain manufacturer evidence that applicable procedures are actually being followed, to show that materials declarations have been assessed, and to check if suppliers can be trusted.

Employment of testing to verify compliance with the requirements of any applicable legislation is usually seen as a last resort. However, in order to obtain conclusive proof of actual product compliance, brands/producers may choose to carry out analytical testing since "Enforcement Authorities" typically also choose to carry out testing to verify the claims of the brands/producers.

7.4 Best Practices

Due to ongoing changes in legislation worldwide, the described task of collecting appropriate supply chain information has become more and more difficult. In addition, new regulatory changes will require having a robust, proactive compliance infrastructure in place. A recent benchmarking study by the Boston based market and benchmarking research organization the Aberdeen Group (see HYPERLINK "http://www.aberdeen.com/" http://

www.aberdeen.com 2006) indicated that compliance performance is less dependent on level of effort than on implementation of best practices and enabling those practices with the appropriate compliance infrastructure. Key recommendations put forward in their study included:

- Adapt proactive compliance strategies, seeking to meet all published standards for current markets, and consider meeting or exceeding strictest global standards in order to enable global sales.
- Proactively monitor and assess compliance early in and throughout the product life cycle, embedding compliance processes into conceptual designs and new product development processes.
- Seek more detailed product composition from suppliers, and in turn target the ability to provide more detailed product disclosures to customers as needed
- Audit content in addition to designing for compliance to address potential variability and data inaccuracy in supply chains.
- Standardize and centralize compliance processes and organizations, leveraging experience and expertise across the enterprise.
- Automate compliance processes with a compliance infrastructure, for repeatability and sustainability—including providing visibility to requirements, documenting product configurations, gathering data from suppliers, assessing compliance, and documenting compliance to support customer documentation on regulatory audits.

Figure 7.1 illustrates a key shift in approach that Intertek strongly supports regarding the use of regulatory/good practice information during the design phase of a new product.

Key activities that should be systematically addressed are listed below and may be expanded into other areas, e.g., safety, quality and performance standards pending upon each company's business requirements.

- Tracking of ongoing changes to global legislation pertinent to restricted/regulated substances as well as good management practices/voluntary standards (e.g., non-government organization's black lists).
- Collection of information on the absence or presence (and if possible and applicable concentration) of *all* substances

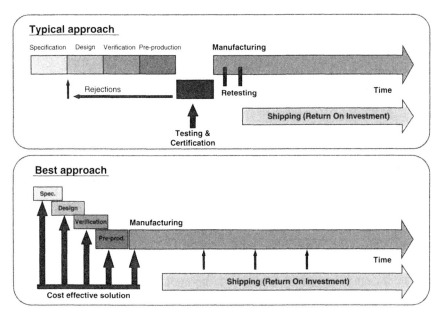

Figure 7.1 Schematic representation of compliance management during a product lifecycle; top: typical current approach; bottom: example recommended approach.

     in products into a centralized database taking into account information flow along full supply chain and role of suppliers/ laboratories.

- Comparison of information on substances in current product to legislation and good management practices/voluntary standards.
- Screening of new product designs to minimize product risk and ensure compliance.
- Storage of key documentation related to restricted substances management.
- Generation of flexible reports related to each company's business requirement.

7.5 Recommendation and Conclusions

Most companies view global compliance as an expense rather than a competitive advantage. Using marketing to promote compliance can appeal to the environmentally conscious customer, consumer, retailer, and distributor, through the use of brochures, web sites, and media communication efforts. If there is a choice between two (or more) similar products, position the environmentally friendly product as the competitive advantage because of its compliance. Sales teams should be trained to use this technique. Examples include the publication of compliance certificates, third party certification marks placed on products or brochures, statements on customs declarations, packaging slips, and shipping boxes. This will also make it easier for enforcement agencies to view compliance without a formal request.

Some businesses even promote their global compliance. These companies provide sustainability reports and report the amounts of recycled hazardous substances. One has even positioned its products as the first ever "green" solution. These businesses are appealing to consumers eco-friendly attitudes while meeting global, regional, or local laws. As Sir T. Leahy, CEO of Tesco one of Great Britain's largest Retailers, stated (*Financial Times Report*, June 12, 2006): "The battle to win customers in the twenty-first century will be increasingly fought not just on value, choice and convenience but on being good neighbours, being active in communities, seizing the environmental challenges, and on behaving responsibly, fairly and honestly in all our actions."

Industries can work together to further promote and comply with restricted substances legislation. Similar to the approach each individual company has taken to comply, the industry should work with suppliers and associations to provide education. These same groups can utilize relationships with lawmakers to lobby for their members best interests. The industry association becomes a central resource for the impacts of new or changing laws on their specific industry and members. Together these will define industry wide strategies.

Recently, a group of manufacturers came together in Japan to form a consortium. Their efforts include setting self-enforced hazardous substance level restrictions in electrical and electronic products which were even more stringent than the European regulations. The same group has also developed a standard format for material declaration each will adhere to.

As demonstrated, compliance is an ongoing process. It may take months or years to fully execute and even then they must be reviewed as components or providers change, as new products are launched, and new laws take effect. Quality control efforts, both for products and systems, must become an ongoing part of everyday business. Environmental directives impact OEM's, suppliers, base material manufacturers, distributors, retailers, customers, governments, and inspectors. It is the responsibility of each business to adhere to all legislation thereby protecting our environment. A strong strategic partnership with a global specialist such as Intertek will ensure an ongoing success of the quality, compliance and safety processes, which are key components of the overall business process enabling sustainable business growth.

8

In Vitro Toxicology for Cosmetics: Regulatory Requirements, Biological Limitations

Ray Boughton

Intertek Testing Services, Leicester, UK

8.1 Introduction

Over the last few decades there has been increasing pressure from a variety of sources to limit and replace the testing of cosmetic products on animals. However, the initial ban on animal testing was postponed on several occasions due to a lack of suitable alternative methods.

More recent changes to the EU Cosmetics Directive have seen the re-introduction of an EU-wide testing ban of finished cosmetics. This is to be followed by bans on the testing of raw materials to be used in cosmetics, and finally with a marketing ban of cosmetics and their ingredients that have been tested on animals to ensure the products safety.

Stringent timelines have been imposed on these prohibitions and much work is currently being carried out in an attempt to devise and refine

C. I. Betton (ed.), Global Regulatory Issues for the Cosmetics Industry Vol. 1, 83–115
© 2007 William Andrew Inc.

alternatives to animal testing. In this chapter, we will look at some of the most important toxicity endpoints, and compare the "classical" animal tests with the currently available (be they approved or simply developmental) *in vitro* methods.

We shall see that for the more simple, point of contact endpoints, *in vitro* testing is generally well correlated with traditional animal test results—and in these circumstance their use is both welcomed and encouraged. But in many cases the *in vitro* methods offer a much poorer breadth and depth of *in vitro* data, and in some cases simply cannot mirror the complexity that exists in vivo. However, we shall hopefully also see that the two approaches can be complimentary.

8.2 EU Cosmetic Legislation and Animal Testing

A number of factors have led to an increased interest in the pursuit and development of *in vitro* alternatives to animal testing. Among these have been both ethical considerations and the requirements of new legislation, which itself may or may not have arisen in part due to ethical considerations or pressure from the public.

In 1986, the EU, in Article 23 of Directive 86/609/EEC[1] on the approximation of laws, regulations and administrative provisions of the Member States regarding the protection of animals used for experimental and other scientific purposes stated:

> The Commission and Member States should encourage research into the development and validation of alternative techniques which could provide the same level of information as that obtained in experiments using animals but which involve fewer animals or which entail less painful procedures, and shall take such other steps as they consider appropriate to encourage research in this field.

The term "alternative" within this legislation is a reference to the classic textbook *"The principles of Humane Experimental Technique,"* released in 1959.[2] This seminal work was part of a project aimed at studying the ethical aspect of laboratory work and was sponsored by the Universities Federation for Animal Welfare. Its key concepts have become known as the 3Rs; *R*efinement in terms of reduced pain or distress, *R*eduction in the number of animals required to perform any given test, and *R*eplacement of animal tests with non-animal alternatives.

Within the field of Cosmetic Safety Assessment it is the last of these (Replacement) that has become of particular importance, as the animal-based assessment of cosmetics has been an area of considerable ethical debate. As a result much *in vitro* sponsorship has been aimed at eliminating animal testing for what is perceived as reasons of vanity. Furthermore, a lot of pressure has been levied by the animal protection communities and the public at large in relation to the testing of cosmetics on animals. To that end there has been a ban on the testing of cosmetics on animals within the UK since 1998.[3]

An EU-wide ban has been more difficult to achieve and has been debated for sometime. The 6th Amendment to Directive 76/768/EEC, published in 1993,[4] introduced the idea of a marketing ban on cosmetic products tested on animals that was to be effective from January 1, 1998. However, this regulation depended on the availability of effective alternative testing methods *at that time* and the ban was subsequently postponed on two occasions.

After some protracted discussion a 7th Amendment to 76/768/EEC[5] was finally published in the first quarter of 2003. This Amendment also includes a prohibition on animal testing. The regulation put into place a ban on the testing of finished cosmetic items within all EU countries as of September 2004. There is then a prohibition of the testing of cosmetic ingredients to be enforced no later than March 2009.

As such it is still therefore legal to sell cosmetics within the EU that have been tested on animals, so long as this testing was itself not conducted within the EU. A slight loophole that will be closed by a marketing ban, prohibiting the sale of finished cosmetic items tested on animals, irrespective of where the testing was performed, as of 2009. There will be an exception to this in that tests for repeated-dose toxicity, reproductive toxicity, and toxicokinetics will be allowed until March 2013.

This exemption reflects, in the main, the fact that alternatives for these test methods are the least far advanced. In either case, the timelines have been clearly established and there will have to be non-animal alternatives available for the safety testing of cosmetics and their ingredients, available within the next 2–6 years.

8.3 A Viable Timeline?

Given this rather stringent timeline one would be forgiven for thinking that *in vitro* alternatives are far advanced and near implementation. In reality, there are still many areas that require a great deal of development.

To some extent this is not as bad as it would at first seem. For example, the 7th Amendment to 76/768/EEC[5] itself points out that "the safety of finished cosmetic products can already be ensured on the basis of knowledge of the safety of the ingredients that they contain."

Indeed, this is what is done on a daily basis when assessing the safety of cosmetics sold in the UK, where the ban on animal testing has been in place for several years. This assessment process does, however, require one important thing; accurate, detailed, and reproducible information on the toxicity of the ingredients.

Toxicologists, and to a lesser extent regulators, define threshold levels and concentrations for substances and preparations. From this an assessment of risk, as a probability in relation to knowledge about intrinsic hazard and predicted in-use exposure, can be made. This is aimed at minimizing the risk to consumers, as the only way in which all risk could be avoided would be to prevent any susceptible individuals being exposed to any hazardous chemical. As individual susceptibility varies greatly, the only way to be 100% risk free would be to prevent any exposure to any chemical, including water!

Obviously this would not be a sensible way to proceed and the fact of the matter is that many chemicals and preparations are only functional for their intended use if they have some toxic properties. Paracelsus' assertion "the right dose differentiates a poison and a remedy"[6] is just as true today as it was in the sixteenth century. It may be slightly amended on technical standpoints to encompass the dose at the target organ, as opposed to the dose administered, but nevertheless the concept of toxicity being based on a combination of intrinsic hazard and exposure must always be borne in mind.

However, there is a general lack of public awareness of these issues and this has become increasingly more obvious in recent years. It is in some respects strange, as our evolutionary history is such that we all continually weigh up risk and benefit in our everyday lives.

Most existing cosmetic ingredients have already been extensively tested on animals, and a wealth of data are available for considering their suitability. On an even more basic level the fact that they have been used for extended periods is good evidence that they are not highly toxic when used in this way.

The major problem, therefore, lies with the development of novel cosmetics and/or ingredients. Without detailed information on the toxicity of an ingredient, one is forced to use the precautionary principle. In other words, if no reliable data are available, one must assume that it is unsafe. If the EU wishes cosmetic companies to remain competitive and innovative, then there has to be robust safety assessments of novel ingredients. Toxicologists can then use these data to support the use (or not) of new ingredients in specific applications.

In practice, developing alternatives for animal test methods is a time consuming process, and can be split into six main phases.[7] The first of these is the research and development (R&D) work involved in generating a new concept; and it is not possible to put a time frame on this, as it may vary greatly on a case-by-case basis.

The other phases of method development can be more easily generalized. The process of prevalidation typically takes at least 2 years, with ensuing validation requiring a further year or more. Peer review and endorsement by the ECVAM (European Centre for the Validation of Alternative Methods) Scientific Advisory Committee (ESAC) takes approximately 1 year. Acceptance into EEC Regulation takes 2 years or more and OECD (Organization for Economic Co-operation and Development) acceptance can take in excess of 5 years. It is only when a method is accepted and published as an OECD Guideline that it becomes truly accepted internationally and only then can global acceptance of data, and its application be assured.

This means that developing an *in vitro* method (once it has undergone the R&D concept phases) can take many years, even if everything goes accordingly—and may take considerably longer if there are problems at any of the stages. By contrast, the regulations state that we need to have methods in place within 2 years.[5] Obviously there are some methods already in stages of validation, but this number is very small when compared to the number of tests and testing regimes that will be required.

8.4 Acute Toxicity

Acute toxicity is one of the most common conventional toxicity tests to be performed. Typically, acute testing involves the application of a single dose of a compound, or in some cases several doses over a 24-hour period. It is also common to test different doses of the chemical on different groups of animals in order to give some information on dose response, i.e., what concentration of the chemical may be safe compared to those that cause death or other types of toxicity.

After administration a number of factors are usually monitored in order to give an indication of toxicity. Amongst the most important of them is the determination of the median lethal dose (LD50), in other words the single dose that will result in the death of 50% of test animals. LD50s are also of value in that they give an overview of the relative toxicities of compounds; and they can be ranked along a spectrum from supertoxic, extremely toxic, highly toxic, moderately toxic, slightly toxic to practically non-toxic.

In addition to gathering information on a dose required to kill organisms outright, examination of autonomic, behavioral, and sensory responses can also be measured along with postmortem analysis. Such information can then be used to help predict likely doses and toxic effects in more long-term studies and the organs that may be of interest.

So in some respects acute toxicity testing can be seen as a reduction and refinement-type test. Though it may use a number of animals and test doses itself, the information generated by it can help to determine what follow-up tests are performed, and helps to ensure that only a few relevant doses are tested. As such it helps refine and reduce the tests and number of animals used in longer-term toxicity tests of that compound.

Moreover, the information on the single dose required to cause toxicity is of great value for a number of different reasons. A useful piece of information, knowing the relative toxicity of a chemical gives information on controls such as handling and storage to minimize risks of accidental exposure to high concentrations of particularly hazardous substances.

From a cosmetic safety standpoint, you can look at what doses do not cause problems compared to those that do. A useful piece of information is the no observed adverse effect level (NOAEL). Using this type of data it may be possible to establish what concentration a human may be safely

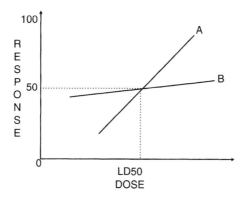

Figure 8.1 Dose–response curve.

exposed to. And it is important not only to know the LD50, but to also look at the dose response. For example, Figure 8.1 shows dose responses for chemical A and B, both with the same LD50. However, it is clear that although at higher doses A is more toxic than B, at lower doses B is more toxic than A.

This information is valuable in not only determining levels that could normally be tolerated by humans (if any), but also in helping to decide which chemicals may be of concern as trace contaminants. Traces of chemical B are more likely to be a concern than trace of chemical A in this example. Restrictions can then be put on these chemicals at either a legislative or safety level to help ensure public safety.

Of course, this assumes that humans have the same susceptibility to the compound as the test animal, which is not always the case. However, by testing a compound in a number of animal species one can look for variations across these, as well as different strains of the same species. By collating this information together one can examine whether an effect may be species specific, or whether it is a more general response.

It can clearly be seen, therefore, that acute toxicity testing is a mainstay of the current animal testing procedures and can provide a wealth of vital information—relating not only to the toxicity of the compound, but also to helping refine and reduce the later stages of toxicity testing. Resultantly finding an accurate and reliable alternative for this test is important in order to ensure safe testing of chemicals going forward.

The most common *in vitro* approach to acute toxicity testing relates to using basal cytotoxicity studies. A common principle of these tests is to measure neutral red uptake (NRU) in cell culture. Neutral red is a dye that readily diffuses into cells, where it is sequestered in lysosomes (acidic organelles that contain digestive enzymes to break down organic matter).

The theory is that toxic compounds inhibit this uptake into lysosomes, by either damaging the cells or their components. One can therefore look for a relative decrease in NRU compared to controls, and use this to give a quantitative ranking of overall toxicity.

However, this standard approach may not always be particularly reliable, as indicated by a recent Interagency Coordinating Committee on the Validation of Alternative Methods (ICCVAM) document.[8] Here it was found that tests with murine fibroblast (3T3) and primary human keratinocyte (NHK) cells produced differences in the IC50 (inhibitory concentration 50%) values obtained between test laboratories for even simple compounds such as sodium chloride.

Likewise there were some chemicals, such as carbon tetrachloride, for which IC50 values could not be determined using the *in vitro* approach. Furthermore, comparing the predicted globally harmonized system (GHS) classification from the *in vitro* studies with those from *in vivo* studies showed a correlation of only 26%.

There are several possible reasons that may account for this and that are potential shortcomings in a variety of *in vitro* tests for different types of toxicity. Foremost amongst these is likely to be considerations of absorption, distribution, metabolism, and excretion (ADME) that occur *in vivo,* but that may not be adequately accounted for *in vitro.*

Some attempts to mimic metabolism have been developed, but these only look at replicating oral toxicity using a liver S9 fraction to imitate the *in vivo* metabolism. Oral administration is but one route of toxicity, and with regards cosmetics probably the least frequently encountered. Acute dermal toxicity is of far greater relevance for cosmetics, but as of yet there have been few efforts to develop an *in vitro* replacement.

This is important because different routes of exposure (dermal compared to oral) can result in widely differing ADME outcomes. The liver is one of the major points of metabolism in the human body; however, the skin also contains a limited number of metabolizing enzymes. Resultantly, metabolism on the skin will be different to metabolism in the liver. Also metabolism in the liver (if the compound ends up there) may be different if the compound is administered orally as opposed to dermally, because of initial metabolism on the skin that does not occur in the stomach, etc.

Similarly, it is not clear how acute inhalation toxicity could be assessed by *in vitro* tests. Furthermore, as with most *in vitro* test methods the compounds of interest have to be highly soluble in water with a limited pH reserve. Materials at the extreme ranges of pH react with culture media and as such cannot be accurately tested. Similarly, hydrophobic materials do not mix with the aqueous test medium and likewise cannot be adequately tested.

A number of practical concerns also exist with regards the NRU protocol specifically. Amongst these are the ability of the NRU to adequately describe the risks associated with volatile chemicals, colored compounds and those chemicals that have specific effects on endosomes (the vesicles formed during the uptake of extraneous cellular material) or lysosomes.

Obviously when talking about cosmetic ingredients we are potentially talking about colors (pigments), compounds with high alkalinity (depilatories) or that are hydrophobic (as the skin is mostly lipid). The current drawbacks of *in vitro* tests therefore apply to chemicals that are likely to be of interest to cosmetics manufacturers. And oral exposure not being the intended route (save for toothpastes, mouthwashes, etc.), it is also going to be necessary to look into developing methods that not only overcome these factors, but that can account for acute dermal or inhalation toxicity.

Even if a test could be devised to accurately give LD50 style results, it does not yield the depth of information on potential target organs and other specific types of toxicity such as carcinogenicity. This will mean that a wider number of follow-up tests will need to be performed in order to correctly identify any possible risks.

Without such basic information as currently offered by animal-based acute toxicity testing it will be very difficult to justify the safety of novel cosmetic ingredients. It is most likely that no single *in vitro* test will be able to provide the same amount of data as studies on living animals. As such there will need to be a battery of different tests used when assessing each aspect of acute toxicity.

Whether this will be successful, and whether there will still be problems due to tests being carried out in isolation without interaction between them, has yet to be seen. At any rate such a battery of alternative tests will take a long time to validate separately and then more time to determine the best mix. Given these problems it would seem unlikely that adequate *in vitro* alternatives will be available for the required 2009 cut-off.

8.5 Skin Corrosion

Skin corrosion is an area in which there has been good progress, and there are already validated methods listed in EU legislation.[9] One of the reasons for this is that the assessments required are not particularly complex; it is simply a case of predicting "on-contact" effect.

In fact to some extent, there is not even the need for *in vitro* testing to predict a corrosive effect. Anything with a pH below 2 or above 11.5 is automatically considered corrosive on the basis of its acidity or alkalinity. In such instances, a simple pH test is sufficient to give an indication of corrosivity.

It is important, however, to note that *in vitro* methods (Rat Skin Transcutaneous Electrical Resistance (TER),[10] EpiDerm,[11] EPISKIN[12]) can be used for performing dose–response assessments. This is extremely useful when considering risk, as a simple corrosive/non-corrosive classification is in itself not all that useful.

Take, for example, a depilatory cream. Generally, these involve the use of sodium hydroxide, a corrosive compound, in order to facilitate the removal of the hair. Because there is good information available on which concentrations of sodium hydroxide give rise to various levels of corrosivity/irritancy, it is possible to determine what concentration of sodium hydroxide can be safely used and under what condition. Without information on the dose response relationship, such judgements would not be possible.

In EU Law, because there are three validated test methods accepted within the EU Dangerous Substance Directive (1999/45/EC),[13] it is therefore not permissible to use *in vivo* tests for corrosivity. Future developments in this field may involve the use of in silico systems, but even if this does not prove effective there are acceptable alternative tests already being used today.

8.6 Skin Irritation

Skin irritancy is a key field in cosmetic safety assessment, with the majority of cosmetics being applied topically. Also, the level of irritancy that is acceptable within a cosmetic product varies depending upon the actual nature of the item and its intended use. For example, with a depilatory product consumers are usually willing to accept a much higher level or possibility of skin irritancy than they would with a face cream.

This means that being able to accurately assess the irritancy of a chemical (and finished product) is not just important from an overall suitability standpoint, but also in relation to how irritant something is at a given concentration. From this information one can then determine whether that level of chemical (and hence irritancy) is likely to be deemed acceptable for a consumer—both in terms of safety and expectancy.

Current animal tests rely on a subjective human scoring system. Compounds are applied topically to the skin of rabbits, and held under a 1-in. gauze for either 4^{14} or 24^{15} hours. At the end of the test period the patch is removed and the skin "scored" for erythema (redness), eschar (scabbing), and oedema (swelling). The skin is then scored again at 72 hours. A second set of tests, using animals with abraded skin can also be done, and this helps show the effect of the compound on damaged skin.

An obvious problem with this is the subjectivity of the scoring, with the possibility of variable results. What one person may perceive to be a certain score, another person may not. This means that the results for any given chemical can vary both between and within labs, essentially dependant on the actual human who is making the judgment.

Though subjective, this system does allow a (rough) grading system of potency. It also provides information on reversibility and long-term damage. The abraded skin test also provides more detail on the lower level irritants, and provides valuable data on what chemicals should be kept away from damaged skin. Obviously with a dermal cosmetic, this kind of information can be crucial in determining the suitability of a compound for use in an eau de toilette, which is applied post abrasion (shaving).

In vitro research and attempts to replace the current animal model have been varied, despite the fact that irritation is a point of contact endpoint. Two main problems contribute to this fact; firstly, that the current *in vivo* method is subjective and so it can be hard to correlate *in vitro* data with variable *in vivo* data; and secondly that irritation itself is not fully understood at the molecular level and is in reality a complex biological response.

In general the *in vitro* alternatives involve the use of cell cultures (usually keratinocytes), and examination of cell viability following application of the test chemical. There are also a number of more complex systems using organotypic cultures or reconstituted models of human skin, though again cell viability is the usually measured parameter to determine irritancy.[16,17]

It is not clear how reliable an indicator of irritancy cell viability is, and a number of other factors (such as production of irritant mediators) have been proposed. Such endpoints are not as thoroughly studied and as of yet have also not been proven fully representative of *in vivo* reactions.

The most promising of these methods are generally the reconstituted models, which rely on growing three-dimensional cultures of skin cells, and currently there are two models (EpiDerm[18] and EPISKIN[19]) that are moving into the later phases of validation. However, these models still present a number of issues that are at basic stages of development and are unlikely to be available in time for the requirement of a full *in vitro* replacement.

For example, the *in vitro* models to do not provide any information about the reversibility of the irritant affect, i.e. how long it will last and how easily the damage is repaired. In addition, while the methods are quite good a distinguishing between non-irritants and irritants, distinguishing moderate irritants from the other two categories is considerably more problematic. Likewise there are some physiological aspects of irritation (such as the effects of inflammatory responses) that are not currently considered by *in vitro* models.

As previously mentioned in cosmetic safety assessment it is often important to know not just pure irritant/non-irritant information, but also to have information of the severity of this irritancy. Many cosmetics (shower gels, bubble baths, etc.) can have quite high concentrations of surfactants (which are often skin irritants) within them. In some cases, the levels present would even cause the product to be labeled as irritant if sold as a general household chemical. However, because of the "rinse-off" nature (i.e., either diluted in use or applied neat but washed off very quickly) of the items the amount of irritant can be considered acceptable.

The "level" of irritancy of a chemical is often therefore critical information, and a system that cannot adequately distinguish moderate irritants from either severe or mild irritants is of limited use. Particularly when considering irritant chemicals that may be used in shaving products, as there will be damaged skin. Under these circumstances it is necessary to have good data on the level of irritancy and ideally the irritancy under that condition.

It is not yet clear how factors such as reversibility or irritancy to abraded skin can be incorporated into the available *in vitro* methodologies. With these concerns, and the difficulty of adequately assigning a compound to a level of irritancy using the existing methods, it is debatable as to whether a suitable alternative will be validated and approved in time for the 2009 cut-off.

8.7 Eye Irritation

The state of progress of *in vitro* tests for eye irritation is, in many ways, similar to that of skin irritation. Again a common problem with developing new methods lays in the comparison of *in vitro* and *in vivo* data being difficult due to the subjective scoring of the *in vivo* method. This further highlights the fact that *in vivo* methods may not always be ideal within themselves, and that the use and development of *in vitro* systems can help to identify the weak aspects of such methods.

An area in which eye irritancy may progress beyond its dermal counterpart is the ease with which one can use isolated eyes from bovine, porcine, avian or lapine origin. The isolated eyes are usually obtained from avian slaughterhouses.

Isolated organ or corneal tests are useful in that a number of parameters (corneal opacity, permeability, hydration, thickness) can be measured. All the layers of the cornea (epithelium, Bowmans membrane, stroma, Descemets membrane, endothelium) can be screened, making it useful in predicting whether a specific chemical affects specific layers. The tests are reasonably straightforward to conduct, have a number of reproducible endpoints and many chemicals can be applied neat, though solid or gaseous substances pose some practical difficulties.

Histological evaluation has been a common endpoint used in *in vivo* studies to determine the depth and degree of injury. The use of ex vivo tissues offers this same opportunity, an important factor as the opacity/permeability measures do not correlate *per se* with degree and depth of injury for some chemicals.[20] Some chemicals, hydrogen peroxide for example, react with the keratinocytes, more than the epithelia, which can lead to a delayed onset of the irritant response.[21] An *ex vivo* test can be used to identify this kind of lesion.

Ex vivo testing also allows for a graded approach to determining toxicity. For chemicals that produce substantial opacity/permeability changes, a direct assessment of a severe irritant can be made. With chemicals that produce limited changes to these parameters, histology can then be used to investigate lesions that may indicate other toxic endpoints.

There are, of course, the inherent methodological differences between *in vivo* and *in vitro* studies. For example, the inflammation that occurs *in vivo* following exposure to mechanical irritants cannot be readily detected *in vitro*. Similarly, there may be strong cytotoxic reactions *in vitro* that may not be evident *in vivo*, depending on solubility and exposure times.

Overall there are a number of reduction/refinement techniques available in the field of eye irritancy testing. However, the validation and acceptance of one or a batch of replacement methods has not yet been achieved. With the work required to finish validating and standardizing individual *in vitro* methods, along with the subsequent development of a batch of test methods that can be routinely used to replace the Draize method,[15] it is unlikely that a full regime of non-animal tests will be in place for the required cut-off enforced by the 7th Amendment of 76/768/EC.[5]

8.8 Skin Sensitization

Skin sensitization is an important component in assessing the safety of cosmetic ingredients and finished products. Aspects relating to it have also come under increased labeling requirements in recent years. The 7th Amendment of 76/768/EC[5] introduced the requirement to label any of the most common 26 human skin sensitizers on product labels if present at more than 0.001% in a leave-on cosmetic.

This requirement takes a blanket approach to simply informing the consumer of the presence of a particular chemical of concern, and does not take into account any notions of potency. Among the 26 chemicals listed some have been shown to cause sensitization at levels much lower than the others. Also the list is far from extensive and there are many more human skin sensitizers in the world today than those 26.

While the labeling is a welcome step to help consumers avoid products that contain a substance they may be allergic to, it is still important to know the potency of sensitizing chemicals in order to determine what levels are safe. Particularly as, unlike irritancy where the response is usually reversible if given enough time, inducing an allergic reaction in an individual results in a life-long concern. Once allergic, even small concentrations of the compound in question can result in quite severe toxic effects.

The latest animal test for determining skin sensitization is the Local Lymph Node Assay (LLNA),[22] which is approved under Annex V of EU Dangerous Substance Directive[13] and also as per OECD 429.[23] It involves the application of a test substance to the mouse ear on 3 consecutive days. On the 6th day, the proliferation of the lymphocytes in the draining lymph node is measured, typically by [3H]-methyl thymidine incorporation into DNA. It is then possible to compare the level of radioactive material with that seen in control animals dosed with vehicle only. A positive identification as a skin sensitizer is when a threefold higher reaction is seen in the test group compared to controls.

The method has been accepted as an alternative to the traditionally used Guinea Pig test methods,[24,25] as it both reduces and refines the use of animal testing for the sensitization endpoint. Moreover, the concentration of a chemical required to produce the threefold higher reaction (the EC3) can be used to provide quantitative measures of the relative potency of chemicals.

Alternatives for sensitization testing are broadly split into the *in silico* Quantitative Structure-Activity Relation (QSAR) methods and *in vitro* testing. Indeed a number of QSAR systems are currently in use in industry for predicting skin sensitization, the most common being DEREK (Deductive Estimation of Risk from Existing Knowledge),[26] TOPKAT (TOxicity

Prediction by Komputer Assisted Technology)[27] and CASE (Computer Automated Structure Evaluation).[28]

DEREK is perhaps the most advanced of these systems and has been in development over the last 15 years. Although these systems have a complex expert rules system that looks at the likelihood of reactions occurring with skin proteins and also considers predicted skin permeability, it is unlikely that they could provide a non-animal alternative in the near future.

The reason for this is the rather obvious fact that they are expert rule systems. As such they only work to the current level of knowledge and require updating whenever new knowledge becomes available. In essence they are simply a computerized version of an expert in the field, with a specific set of knowledge. They may fail with novel, untested chemical classes because without existing toxicity data and knowledge, they cannot function.

Of equal concern is the fact that available *in vitro* methods are currently rather unreliable. Foremost amongst the problems is the fact that allergic contact dermatitis (ACD) is a complex cell-mediated immune response, which is still not fully understood from a mechanistic standpoint. Basic requirements for ACD are an ability to penetrate the skin, react with protein and to then be recognized as antigenic by the immune system. It is not clear how an *in vitro* method can be designed to take into account all of these factors. As a result most *in vitro* tests look at only one aspect of this process.

A commonly used method is to culture keratinocytes and then treat them with various chemicals. Post exposure the excretion of a number of markers (IL-1,[29] IL-8,[30] CD40,[31] CD80[32]) can be monitored. So far, no marker has proved reliable and the response observed seems to be chemical specific, as opposed to a general sensitizers vs. non-sensitizers response.

Ideally Langerhans cells would be used in testing because, although they constitute a relatively small percentage of the total skin cells, they are the mediators of the allergic response. Langerhans cells are notoriously difficult to extract in large numbers, as well as being difficult to successfully culture and direct work on them has therefore proved problematic. One alternative has been to use dendritic cells from peripheral blood,[33] as they share a number of similarities to Langerhans cells, but are easier to extract and culture.

However, some recent developments have proved promising. For example, there are noticeable differences in the internalization (as measured by using fluorescent labeling) of the cell surface receptor HLA-DR when cells are exposed to either irritants or sensitizers.[34] Irritants tend to result in HLA-DR grouping in small vesicles with a diffuse fluorescence, while sensitizers cause HLA-DR to group in larger vesicles with a brighter fluorescence. Similarly sensitizers resulted in HLA-DR being internalized to lysosomes that collect near the nucleus, whereas non-sensitizers did not.

It is not clear at this stage whether this effect is seen in wide range of sensitizers and non-sensitizers as the current tests have involved only a limited number of compounds. Furthermore, it is not immediately evident how these observations would allow the relative potencies of various allergens to be determined.

Co-culture and organotypic models are, by and large, similar to single cell cultures and share many of the same drawbacks. *In vivo* the induction of a sensitization response involves the interaction between Langerhans cells and T-lymphocytes (T-Cells). Therefore, a culture system involving both of these cell types would appear to provide a sensible approach to *in vitro* testing. Typically, this involves the use of antigen-modified Langerhans or other dendritic cells, to promote increased proliferation of T-Cells or T-helper cells. Initially this method showed some promise but there have been problems in determining moderate and weak sensitizers.

The inability of *in vitro* models to accurately rank the potency of sensitizers is a major concern, because such data provides a valuable indication of what levels may be safe for human exposure. In the cosmetics industry this is particularly important, as many of the components used in the manufacture of fragrances are human skin sensitizers. Likewise many essential oils (as natural products are currently an important trend) contain high levels of potential skin sensitizers. Orange oil, for example, can contain up to 95% Limonene. Knowing how potent these compounds are and setting suitable safe exposure limits is therefore a large part of the assessment of safety of leave-on cosmetic products.

There are no currently validated *in vitro* alternatives to the LLNA, and there are numerous problems and deficiencies with the current generation of tests. Developing alternative methods to look at just some aspects of sensitization is likely to be lengthy, let alone validating a full testing battery to replace the animal methods. Again this is probably going to fall well outside the timeline of the EU Cosmetics Directive.[5]

8.9 Carcinogenicity

Carcinogenicity is a toxicity endpoint of high concern, with the effects being potentially fatal and irreversible. Similarly effects may not be immediate (a consumer would not have obvious signs of effect, unlike the reddening in skin irritancy for example) and can potentially take years to develop. Resultantly, it is crucial to adequately protect consumers from chemicals posing this kind of risk.

Currently, the classification of a compound as carcinogenic is done on the basis of both human and animal data. In the case of human data, this almost always comes from industrial or accidental exposure, and indeed was how the early human carcinogens were identified. Animal studies typically involve long-term studies (over 2 years or more), whereby the numbers and types of tumors are monitored in test animals compared to controls.

The long-term nature of the studies required, means that they cannot readily be mimicked by short-term *in vitro* tests. While animal studies themselves are not ideal (there can be significant inter-species differences in the response to carcinogens) the seriousness of it is such that it would not be sensible to simply wait for accidental human exposures in order to classify something.

The reason that carcinogenicity can vary between species is that it's a multistage process involving a sequence of complex biological interactions, often taking place over many years. In addition, chemical carcinogenicity involves a wide variety of target organs and tissues. The site of tumor formation can depend on how the chemical is administered and subsequently absorbed, metabolized and distributed within the body as well as how (and if) it is excreted. As all of these factors can vary widely between species, it is natural that the outcome (carcinogenicity) also varies.

While this makes extrapolation of animal data to humans difficult, testing in a number of different species and the ability to identify the organs effected and why, helps us to get a broader picture on whether the effect may be relevant to man or a species specific concern. Resultantly, it is often possible to classify the risk posed to man, based on the existing test regimes.

A large number of *in vitro* genotoxicity tests have been developed to predict potential carcinogenicity, based on the observation that the carcinogenic process involves changes in the genome. These include gene mutation assays in bacteria,[35] yeast,[36] and mammalian cells[37]; DNA damage and repair,[38] chromosomal aberration[39] and sister chromatid exchange[40] tests in mammalian cells; and mitotic recombination assays in yeast.[41] Versions of all of these tests are approved OECD methodologies as well as being incorporated in Annex V of the EU Dangerous Substance Directive.[13]

Other tests have been developed to investigate tumor progression, such as cell transformation[42] and the gap junction intercellular communication (GJIC) assays to detect promoters and non-genotoxic carcinogens[43]. Neither of these concepts has yet to be fully validated.

While genotoxicity assays have for many years been used to predict potential carcinogenicity, they do not give any measure of potency or indication of target tissues. Also there are a number of cases of positive findings with non-genotoxic substances and examples of non-genotoxic substances that are in fact carcinogenic. It may be possible that some of these will be detectable in the future (e.g., using GJIC assay), but there is insufficient evidence at this stage to conclude so with certainty.

In consequence, because of the complex *in vivo* responses involved, it is highly unlikely that a single *in vitro* test will ever be a substitute for a lifetime animal study of carcinogenicity. The best that can be hoped for at this stage, is that a battery of *in vitro* tests may enable investigators to set priorities for carrying out any animal tests required.

As with most *in vitro* toxicological studies, it is relatively easy to identify strong positive and strong negative responses; it is the gradations of response and the interpretation of the data from these studies that are the most challenging in terms of protection of human health. It is likely to remain thus in terms of assessing carcinogenicity for the foreseeable future.

8.10 Reproductive/Developmental Toxicity

As with carcinogenicity, developing an *in vitro* test for reproductive toxicity is an important field (because of the severity of the effects that may not be immediately evident), but difficult because the *in vivo* mechanisms are complex and in many cases not well understood. Similarly, there is a large range of different endpoints and small changes in the timings of exposure to compounds can produce significant changes in toxicity.

The case of thalidomide is a prime example of this—where it can be taken at most times during pregnancy with little effect, yet when taken at a particular stage causes severe effects on the limb bud formation.[44] This makes assessing reproductive and developmental toxicity very difficult, even in animal studies, as a vast range of different tests at different times in the developmental and reproductive pathways need investigating.

From a consumer safety viewpoint, accurate determination of any reproductive and developmental toxicities are important. Cosmetic products are extensive in range and can be used by the whole cross-section of the public; from childhood to adolescence and adulthood, by pregnant mothers, and couples hoping to conceive. It is of particular concern that it is not even possible to label some items as unsuitable for pregnant mothers, as there is often a period of time at early conception that one is unaware they are pregnant.

Because of these concerns, and the relatively large number of animals required to complete reproductive/developmental testing, this field was one of the first to receive a large amount of EU funding. It was also chosen as a pilot in the development of a series of *in vitro* alternatives to replace a systemic *in vivo* effect. The resulting project (ReProTect) looks to break down the whole reproductive cycle and has picked three major areas to look at; fertilization, implantation, and prenatal development.[45]

An aspect that is immediately lacking is that of post-natal development and behavior. One obvious example is reproductive behavior, which includes such patterns as the establishment of mating systems, courtship, sexual behavior, parturition, and the care of young. Obviously the determination of correct reproductive behavior can only be done by observing living organisms; it is not clear how any *in vitro* test would ever be able to replace this.

Likewise it is not obvious how other complex behaviors that are currently investigated (such as surface righting, fore- and hind-limb placement, auditory startling, forelimb gripping and hunting behavior[46]) could be assessed using *in vitro* methods. While such effects are not directly related to mating process, dysfunction in these areas can result in decreased survivability and ensuing decrease in reproductive output due to shortened lifespan.

Even when considering the development of organs and limbs, it is important to note that this is not complete when offspring are born. Developmental testing *in vivo* can typically involve the study of post-birth factors, required for successful adult development and reproduction. For example, a number of post-natal developmental landmarks can be monitored *in vivo* (day of first oestrus, testicle descent, eye opening, incisor eruption, etc.[46]), and again it is not clear how such factors (that may be not become evident until a long time after birth) could be adequately studied *in vitro*.

Another important consideration in developmental toxicity specifically, is that infants do not develop independently. There is the genotype, health and all other aspects of the female in which the child develops that must be taken into account. This could, for example, affect things such as the rate of and preferred pathway for biotransformation—having a knock on effect on the concentration in maternal blood and so on.

Realistically it is not evident how all of these concerns can be overcome within the next 6 years in order to meet the regulatory requirements of the EU Cosmetics Directive.

8.11 Toxicokinetics

Toxicokinetics is a term used to describe data on the concentration of a specific compound within the various organs and compartments of an organism. It incorporates information such as the absorption of a chemical, how it is distributed, the metabolic pathways involved in detoxification/activation as well as elimination. It should also be noted that all of these factors require investigation for the various metabolites of a compound.[44]

Though the areas it covers can therefore be quite broad, it is often simplified to consider the basic processes of ADME. These processes are behind many important aspects of toxicity and play a role in nearly all of the types of toxicity outlined above—determining the nature and extent of toxicity once a compound enters the body. As the process are often complex, simultaneous and critical to the outcome of a chemicals toxicity, it is of vital importance to ensure these behaviors can be modeled *in vitro*.

Particularly because *in vitro* studies often involve the application of a compound to a test setup that looks at one particular toxicity endpoint in isolation. In order to ensure that this endpoint is valid it is necessary to determine whether ADME allows for the endpoint, and whether the effects of ADME would result in an increased or decreased toxicity *in vivo*.

This is not to say that existing *in vivo* studies are ideal, the complex nature of its effects means that in many cases the results of *in vivo* testing can depend upon the species of animal the tests are performed on. In some respects this can make extrapolation to humans difficult and, whether humans possess the same ADME pathways as the test species, must always be considered. Similar to other aspects of toxicology, testing a compound in a number of different species (or those species most similar to humans) helps determine which effects are species specific and which may be of concern from a consumer (human) viewpoint.

8.11.1 Absorption

Good progress has been made in modeling absorption following dermal administration, despite the fact that it is not usually required for chemicals under the EU Dangerous Substances Legislation.[13] This is an extremely important field for cosmetics, whereby the main mode of application is dermal. There is an approved OECD guideline available for the use of excised porcine or human skin.[47] As this method is already approved it is not expected that further work in the field of *in vitro* dermal absorption will be conducted.

One point of interest, however, is that at the moment no models exist to account for repeated topical dosing or for topical dosing and metabolism combined. As cosmetics can be applied numerous times a day (potentially every day of the year) research into repeated dosing would be beneficial. Likewise the ability to study dermal metabolism in conjunction with dermal absorption could be critical. In some cases, the skin may metabolize a harmful compound to a non-harmful one, or vice-versa, meaning that at present the *in vitro* model would not account for this.

In vitro oral absorption studies are typically done using either artificial membranes designed to mimic the properties of the various components of the gastrointestinal (GI) tract[48]; or by using cell or organotypic cultures of the GI tract.[49] These methods are well accepted in drug research, though there are still a number of concerns with these types of methods. The use of artificial membranes particularly allows only for the study of absorption via passive diffusion.

While this is the main route of absorption for many chemicals, there are three other routes (filtration through membrane pores, carrier-mediated uptake and phagocytosis (engulfing)) that can play a role *in vivo*. Inability to model these factors could be crucially important for some chemicals, leading to an underestimation of absorption and hence toxicity.

Similar problems can occur with using cell culture, as generally this involves the use of cancer cell lines. These lines may not display the standard morphology seen *in vivo*, lacking the ability to produce mucous or having different abilities to mimic the absorption routes mentioned above.

Oral routes of exposure are not common in cosmetics, and in some respects it is the fields of dermal or inhalation exposure that are more crucial. However, the ability to correctly determine absorption through this route will have an important effect on the ability to develop accurate replacements for some of the short-term toxicity and other general toxicity studies.

Determining absorption under inhalation exposure is also frequently done using cell cultures, generally obtained from lung carcinoma tissues.[50] A number of different alternative cell lines exist, with advantages and disadvantages that are in general similar to the problems associated with using cell cultures for oral absorption. Some studies have also been designed to use primary human cell cultures; however, a major drawback in doing this is the ability to obtain such samples for use.

A feature of inhalation absorption/toxicity of specific concern is the effect of particulate matter. *In vivo* a number of responses such as inflammation and ensuing silicosis, etc. may be identified. It is not clear how an *in vitro* system modeling absorption across a barrier could be used to show this effect. Likewise there are factors such as thrombosis and respiratory allergy that may also prove difficult to account for *in vitro*.

Overall absorption studies are well advanced for the dermal route, though more work is desirable regarding the effect of metabolism and repeated dosing. The oral and inhalation routes are less advanced at this stage and it is likely that more focussed work in these areas will be required to ensure appropriate methods are available for the regulatory deadline.

8.11.2 Distribution

In vitro modeling of distribution is currently very limited, with only a few methods available to look at factors such as plasma protein binding (PPB)[51] and tissue partitioning.[52] PPB can be important in toxicology, as in general it is only the fraction of a chemical not bound to plasma protein that is available to cells and tissues. Similarly, the tissue-blood partition co-efficient (the measure of how easily a compound moves into a tissue from the blood) is important in determining how much of a chemical ends up in a tissue and how quickly it happens.

Currently, the few available alternatives rely on incubating the compound of concern with either human plasma serum (for PPB) or with various blood and tissue buffers (tissue-blood partition coefficient) *in vitro* and measuring levels of free and bound chemical. Such methods are relatively simple, and can be done quickly in high volume. In both cases, however, the methods have been developed with a view to pharmaceutical compounds and it is not clear how directly relevant to the cosmetics industry they will be.

In addition, the PPB methods do not currently distinguish between the various plasma proteins to which a chemical may bind, and this makes it difficult to assess when a chemical will become saturated within the plasma *in vivo*. The tissue-blood partitioning also only accounts for what happens in general terms, and does not take account of specific transport mechanisms (such as active transport or carrier mediated uptake) that may lead to increased concentrations within the tissue *in vivo*.

At this stage there are severe deficiencies in the area of chemical distribution and accumulation within the various tissues of the body. Given that toxic effects rely upon the concentration of a chemical at any given tissue or organ, the importance of this field cannot be overemphasized. If we are to have a reliable set of *in vitro* toxicity tests, it is crucial that the concentration of a chemical at each of the tissues/organs can be accurately predicted. A significant amount of work in this field will be required to ensure test regimes are in place by the 2013 deadline.

8.11.3 Metabolism

The importance of metabolism to *in vitro* testing is widely recognized and a recent review document from ECVAM is available on the subject.[53] For the sake of brevity we shall not discuss this in detail here, and suffice with a summary of the current concerns in this field. Mostly these relate to accurately modeling human *in vivo* metabolism and selecting the appropriate metabolizing system for a compound. However, there are also technical barriers to overcome, such as maintaining organ and tissue specific functioning *in vitro*, keeping the metabolizing system(s) stable over prolonged periods *in vitro* and the possible inactivation of these systems by test solvents, etc.

It should also be noted that even in the existing *in vitro* systems that use metabolizing systems, this is often done either prior to exposure (so that the chemical is incubated with a metabolizing system before application) or during exposure but extra-cellularly. The concern here is that *in vivo* metabolism could occur intra-cellularly, meaning that resulting metabolites do not need to cross the cellular membranes and be absorbed. It is foreseeable that a compound could be metabolized to a polar, but highly reactive metabolite; its toxicity may then not be predicted *in vitro* because it is not bioavailable to the test culture.

By using *in silico* computer modeling many of these practical concerns can be overcome, though the *in silico* models themselves are often limited. Computer models can be split into three broad categories; those based on rules around structure and metabolism, QSAR models, and metabolizing protein models.[54] Of the *in silico* approaches, Meteor is perhaps the most well known, and falls under the first category of using structure and metabolism rules to predict the biotransformation of compounds.

As with the use of *in silico* methods in other fields, the main drawbacks are associated with the fact that prediction is usually only accurate for classes of chemicals whose metabolism is already well known. With metabolism in particular there are often concerns relating to the fact that multiple possible routes or sites of metabolism could exist. Predicting which of these will be dominant *in vivo*, and hence what the major metabolite of a chemical will be, is particularly difficult with the current level of knowledge and systems.

Given the importance of this factor (and of toxicokinetic factors in general) to the final toxicity of a chemical, research in this field will be vital to ensuring that a suitable set of *in vitro* replacement tests is available. It is good that the importance of metabolism has been widely recognized and that there are sustained efforts to develop and refine the available *in vitro* methods. Although (at the time of writing) there are 6 years before toxicokinetic replacements are required, its important to note that the widespread impact of metabolism on all areas of *in vitro* replacement means that it is necessary to ensure advancement as soon as possible. It is debatable whether significant enough steps can be taken in the next few years.

8.11.4 Excretion

The final ADME factor of excretion is currently very poorly represented *in vitro*. Some efforts have been made regarding renal and biliary excretion, though these are aimed more specifically at pharmaceuticals and these routes of excretion are less important for cosmetic products. More worrying is the fact that at present there seems to be little effort focused on this aspect of *in vitro* toxicology testing.

As with all areas of ADME, excretion can play a vital role in the final toxicity of a chemical. For example, rapid and efficient elimination of a compound from a tissue results in a lower (or no) concentration of that compound at the tissue or organ. This in turn will result in limited or no toxicity.

Given the lack of existing models and current impetus it is likely that developing a series of tests to deal with Excretion will take well beyond the 2013 deadline.

8.11.5 Toxicokinetics

The uptake and distribution of a chemical within an organism, along with its metabolism and excretion often occurs in a competitive and simultaneous manner. Resultantly what is happening in one organ/tissue can effect what will happen in another. To some extent this makes it difficult to see how *in vitro* tests considering each in isolation could provide the full picture.

Problems can escalate if one has to test a complex mixture of substances (such as a finished cosmetic product) as different component chemicals may utilize different pathways of ADME within the body. With the above-mentioned concerns relating to toxicokinetics and its widespread impact in all other areas of *in vitro* testing it is likely to be a considerable time before all aspects of this subject can be fully accounted for.

To that end this area has the joint longest time until regulatory necessity (2013). Whether it will be possible to develop adequate mechanisms during this time frame is debatable. Moreover, the fact that toxicokinetic factors play a role in virtually all areas of toxicology (including those required by 2009) means that despite the 2013 deadline, good progress needs to be made in the next 2 years.

8.12 Other Considerations

This, of course, covers only some of the fields of toxicology and hopefully gives an overview of the current situation with regards the areas that are of greatest concern to cosmetics. There are a number of other fields that are currently being investigated, such as sub-acute and sub-chronic toxicity, ecotoxicity, and phototoxicity, which will also have an impact on cosmetic testing. These fields, and others, either share similar problems to, or tie into, the various categories above.

Another important aspect that could be of concern to all areas of *in vitro* toxicology is the concept of multifactorial causation. This basically refers to the fact while some effects can be caused by purely genetic factors (Down's syndrome) or purely chemical/environmental factors (high dose X-ray exposure), the vast majority of effects are caused by a combination of the specific insult and the specific genome.

Whether the use of genetically identical, human cells in culture is better or worse than the use of outbred, genetically variable, but non-human animals will be an interesting area for debate.

8.13 Conclusions

As someone who completed their toxicology training in the early part of this century I was very much educated with *in vitro* toxicology. To look at general toxicity we were taught to use cell cultures and measure responses. Likewise if you wanted to investigate carcinogenicity you looked at tests such as chromosomal abberations, Ames assay, adduct formation—and so the list goes on.

The field I currently work in (risk assessment of household and industrial chemicals) means that everyday we have to look at and review the available animal and non-animal data on chemicals. It is important not just to have data, but also to have data that is useable, reliable and meaningful; information on dose response, on mechanism of action, of as many of the things that allow one to correctly determine whether a given substance in a given situation is going to be safe or not.

A major problem we currently face today is that a lot of people (from the public to regulators and even scientists) are looking to entirely replace animal testing with *in vitro* assays. While this may be an admirable long-term goal, it should be kept in mind that it is just that. Given the current level of progress and quality of information generated compared to the existing animal methods it is clear that in many fields *in vitro* methods will simply be unable to compete with existing *in vivo* tests.

Education is sometimes referred to as lying to children. In many ways *in vitro* toxicology can, at this stage, be considered in a similar vein. It provides a simplistic and uncomplicated way to look at things and as such can be useful for pre-screening and generating basic data. It gives quick and simple data that can readily be understood.

But it must be remembered that, for the immediate future, that is all it is likely to be. When more detailed information is required, and it often is, to enable accurate and reliable estimation of human risk, such an approach simply does not work. It is, of course, important to look at how we can develop and move forward in both fields. And if and when possible we should look to move away from animal testing. However, this is likely to be a long road, and one that is quite rocky.

In vitro methods, and particularly *in silico* models, rely on replicating what is currently known. There is no way to know if and how they will cope with the unknown. If we move too quickly and switch to *in vitro* testing when we are not ready, then it is more than likely there will be a few instances of chemicals passing *in vitro* screens only for it to be later discovered, possibly at the cost of some peoples health, that they are not safe and work by a mechanism that could not be predicted by an *in vitro* test system designed around current knowledge.

One of the fundamental truths within toxicology is that no substance or product is ever completely non-hazardous. At certain doses virtually all chemicals have a toxic effect, but at others may be safe for use.

Toxicity is crucially based on a vast number of different variables that are constantly, or potentially constantly, changing. Thus, it must be ensured the information available to safety assessors and regulators is as detailed and as in-depth as possible, so that the greatest public safety can be ensured.

Absolute safety of any cosmetic product or ingredient is not possible, but what we must be mindful of is forcing upon ourselves a paucity of data by insisting that non-animal methods are used when no suitable non-animal alternatives are available. Doing so compromises the ability of companies to develop novel products and ingredients, the ability of regulators and assessors to ensure safety and risks the health of the end consumer.

References

1. EC (1986). Council Directive 86/609/EEC of 24 November 1986 on the approximation of laws, regulations and administrative provisions of the Member States regarding the protection of animals used for experimental and other scientific purposes. *Official Journal of the European Communities*, L358, 1–59.
2. Russell, W.M.S., Burch, R.L. (1959). *The Principles of Humane Experimental Technique*. London: Metheun.
3. SI 1996 No. 2925. (1996). *Consumer Protection: The Cosmetic Products (Safety) Regulations 1996*. London: HMSO.
4. EC (1993) Council Directive 93/35/EEC of 14 June 1993 amending for the sixth time Directive 76/768/EEC on the approximation of the laws of the Member States relating to cosmetic products. *Official Journal of the European Communities* L151, 32–3.
5. EU (2003). Directive 2003/15/EC of the European Parliament and of the Council of 27 February 2003 amending Council Directive 76/768/EEC on the approximation of the laws of the Member States relating to cosmetic products. *Official Journal of the European Union* L66, 26–35.
6. Paracelsus, as cited in Lu, F.C., Kacew, S. (2002). *Lu's Basic Toxicology* (4th edn). London: Taylor & Francis.
7. Manou, I., Eskes, C., de Silva, O., Renner, G., and Zuang, V. (2005). EC-VAM. Alternative (non-animal) methods for cosmetics testing. Current status and future prospects. Chapter 2: Safety data requirements for the purposes of the Cosmetics Directive. *Alternatives to Laboratory Animals*, 33, 21–26.
8. NIH Publication No. 07-4519. (2006). *ICCVAM/NICEATM Peer Review Panel Report: The Use of In Vitro Basal Cytotoxicity Test Methods for Estimating Starting Doses for Acute Oral Systemic Toxicity Testing*. Available at http://iccvam.niehs.nih.gov/methods/acutetox/inv_nru_announce.htm
9. EU (2000). Commission Directive 2000/33/EC of 25 April 2000 adapting to technical progress for the 27th time Council Directive 67/548/EEC on the approximation of laws, regulations and administrative provisions relating to the classification, packaging and labelling of dangerous substances. *Official Journal of the European Communities*, L136, 90–107.
10. Balls, M., Corcelle, G. (1998). ECVAM. Statement on the scientific validity of the rat skin transcutaneous electrical resistance (TER) test (an

in vitro test for skin corrosivity). *Alternatives to Laboratory Animals*, 26, 275–277.

11. Balls, M., Hellsten, E. (2000). ECVAM. Statement on the scientific validity of Epiderm™ human skin model for skin corrosivity testing. *Alternatives to Laboratory Animals*, 28, 365–366.

12. Balls, M., Corcelle, G. (1998). ECVAM. Statement on the scientific validity of the EPISKIN™ test (an *in vitro* test for skin corrosivity). *Alternatives to Laboratory Animals*, 26, 277–280.

13. EEC (1967). Council Directive 67/548/EEC of the 27th June 1967 on the approximation of laws, regulations and administrative provisions relating to the classification, packaging and labelling of dangerous substances. *Official Journal of the European Economic Community*, 196, 1–98.

14. OECD (2002). *OECD Guidelines for the Testing of Chemicals No. 404: Acute Dermal Irritation/Corrosion*. Paris: OECD.

15. Draize, J.H., Woodard, G., Calvery, H.O. (1944). Methods for the study of irritation and toxicity of substances applied topically to the skin and mucous membranes. *Journal of Pharmacology and Experimental Therapeutics*, 82, 377–390.

16. van de Sandt, J., Roguet, R., Cohen, C., Esdaile, D., Ponec, M., Corsin, E., Barker, C., Fusenig, N., Liebsch, M., Benford, D., de Brugerolle de Fraissinette, A., Fartasch, M. (1999). The use of keratinocytes and human skin models for predicting skin irritation. *Alternatives to Laboratory Animals*, 27, 723–743.

17. Zuang, V., Balls, M., Bothan, P.A., Coquette, A., Corsini, E., Curren, R.D., Elliott, G.R., Fentem, J.H., Heylings, J.R., Liebsch, M., Medina, J., Roguet, R., van de Sandt, J.J.M., Wiemann, C., Worth, A.P. (2002). Follow-up to the ECVAM prevalidation study on *in vitro* tests for acute skin irritation. *Alternatives to Laboratory Animals*, 30, 109–129.

18. Cannon, C.L., Neal, P.J., Southee, J.A., Kubilus, J., Klausner, M. (1994). New epidermal model for dermal irritancy testing. *Toxicology In Vitro*, 8, 889–891.

19. Roguet, R., Cohen, C., Dossou, K.G., Rougier, A. (1994). Episkin, a reconstituted human epidermis for assessing *in vitro* the irritancy of topically applied compounds. *Toxicology In Vitro*, 8, 49–59.

20. Curren, R., Evans, M., Raabe, H., Hobson, T., Harbell, J. (1999). Optimisation of the bovine corneal opacity and permeability assay: Histopathology aids understanding of the EC/HO false negative materials. *Alternatives to Laboratory Animals*, 27, 344.

21. Maurer, J.K., Molai, A., Parker, R.D., Li, L., Carr, G.J., Petroll, M.W., Cavanagh, D.H., Jester, J.V., (2001). Pathology of ocular irritation of bleaching agents in the rabbit low-volume eye test. *Toxicological Pathology*, 29, 308–319.

22. Kimber, I., Dearman, R.J., Basketter, D.A., Ryan, C.A., Gerberick, G.F. (2002). The local lymph node assay: past, present and future. *Contact Dermatitis*, 47, 315–328.

23. OECD (2002). *OECD Guidelines for the Testing of Chemicals No. 429: Skin Sensitisation: Local Lymph Node Assay*. Paris: OECD.

24. Magnusson, B., Kligman, A.M. (1969). The identification of contact allergens by animal assay. The guinea pig maximisation test. *Journal of Investigative Dermatology*, 52, 268–276.
25. Beuhler, E.V. (1965). Delayed contact hypersensitivity in the guinea pig. *Archives of Dermatology*, 91, 171–177.
26. Ridings, J.E., Barratt, M.D., Cary, R., Earnshaw, C.G., Eggington, C.E., Ellis, M.K., Judson, P.N., Langowski, J.J., Marchant, C.A., Payne, M.P., Watson, W.P., Yih, T.F. (1996). Computer prediction of possible toxic action from chemical structure: an update on the DEREK system. *Toxicology*, 106, 267–279.
27. Enslein, K., Gombar, V.K., Blake, B.W., Maibach, H.I., Hostynek, J.J., Sigman, C.C., Bagheri, D. (1997). A quantitative structure-activity relationship model for the dermal sensitisation guinea pig maximisation assay. *Food and Chemical Toxicology*, 35, 1091–1098.
28. Gealy, R., Graham, C., Sussman, N.B., Macina, O.T., Rosenkranz, H.S., Karol, M.H. (1996). Evaluating clinical case report data for SAR modelling of allergic contact dermatitis. *Human and Experimental Toxicology*, 15, 489–493.
29. Pastore, S., Shivji, G.M., Kondo, S., Kono, T., McKenzie, R.C., Segal, L., Somers, D., Saunder, D.N. (1995). Effects of contact sensitisers neomycin sulfate, benzocaine and 2,4-dinitrobenzene 1-sulfonate, sodium salt on viability, membrane integrity and and IL-1α mRNA expression of cultured normal human keratinocytes. *Food and Chemical Toxicology*, 33, 57–68.
30. Wilmer, J.L., Burelson, F.G., Kayama, F., Kanno, J., Luster, M.I. (1994). Cytokine induction in human epidermal keratinocytes exposed to contact irritants and its relation to chemical-induced inflammation in mouse skin. *Journal of Investigative Dermatology*, 102, 915–922.
31. Coutant, K.D., Ulrich, P., Thomas, H., Cordier, A., de Brugerolle de Fraissinette, A. (1999). Early changes in murine epidermal cell phenotype by contact sensitizers. *Toxicological Sciences*, 48, 74–81.
32. Wakem, P., Burns, R.P., Ramirez, F., Zlotnick, D., Ferbel, B., Haidaris, C.G., Gaspari, A.A. (2000). Allergens and irritants transcriptionally upregulate CD80 gene expression in human keratinocytes. *Journal of Investigative Dermatology*, 114, 1085–1092.
33. Degwert, J., Steckel, F., Hoppe, U., Kligman, H. (1997). *In vitro* model for contact sensitisation. I. Stimulatory capacities of human blood-derived dendritic cells and their phenotypical alterations in the presence of contact sensitizers. *Toxicology In Vitro*, 11, 613–618.
34. Rizova, H., Carayon, P., Barbier, A., Lacheretz, F., Dubertret, L., Michel, L. (1999). Contact allergens, but not irritants, alter receptor mediated endocytosis by human epidermal Langerhans cells. *British Journal of Dermatology*, 140, 200–209.
35. OECD (1997). *OECD Guideline for Testing of Chemicals No. 471: Bacterial Reverse Mutation Test*. Paris: OECD.
36. OECD (1986). *OECD Guideline for Testing of Chemicals No. 480: Genetic Toxicology: Saccharomyces cerevisiae, Gene Mutation Assay*. Paris:OECD.

37. OECD (1997). *OECD Guideline for Testing of Chemicals No. 476: In vitro Mammalian Cell Gene Mutation Test*. Paris: OECD.
38. OECD (1986). *OECD Guideline for Testing of Chemicals No. 482: Genetic Toxicology: DNA Damage and Repair, Unscheduled DNA Synthesis in Mammalian Cells In Vitro*. Paris: OECD.
39. OECD (1997). *OECD Guideline for Testing of Chemicals No. 473: In Vitro Mammalian Chromosomal Aberration Test*. Paris: OECD.
40. OECD (1986). *OECD Guideline for Testing of Chemicals No. 479: Genetic Toxicology: In Vitro Sister Chromatid Exchange Assay in Mammalian Cells*. Paris: OECD.
41. OECD (1986). *OECD Guideline for Testing of Chemicals No. 481: Genetic Toxicology: Saacharomyces cerevisiae, Miotic Recombination Assay*. Paris: OECD.
42. Isfort, R.J., Kerckaert, G.A., LeBoeuf, R.A. (1996). Comparison of the standard and reduced pH Syrian Hamster Embryonic (SHE) cell *in vitro* transformation assays in predicting the carcinogenic potential of chemicals. *Mutation Research,* 356, 11–63.
43. Rivedal, E., Mikalson, S.O., Sanner, T. (2000). Morphological transformation and effect on gap junction intercellular communication in Syrian hamster embryo cells as screening tests for carcinogens devoid of mutational activity. *Toxicology In Vitro*, 14, 185–192.
44. Boelsterli, U.A. (2003). *Mechanistic Toxicology*. London: Taylor & Francis, pp. 8–12.
45. http://www.reprotect.eu
46. Parker, R.M., Hood, R.D. (2006). *Developmental and Reproductive Toxicology: A Practical Approach* (2nd edn). Boca Ranton: CRC Press.
47. OECD (2004). *OECD Guidelines for the Testing of Chemicals No. 428: Skin Absorption: In Vitro Method*. Paris: OECD.
48. Kansy, M., Senner, F., Gubernator, K. (1998). Physicochemical high throughput screening: parallel artificial membrane permeation assay in the description of passive absorption processes. *Journal of Medical Chemistry,* 41, 1007–1010.
49. Le Ferrec, E., Chesne, C., Artusson, P., Brayden, D., Fabre, G., Gires, P., Guillou, F., Rousset, M., Rubas, W., Scarino, M.L. (2001). *In vitro* models of the intestinal barrier. The report and recommendations of ECVAM Workshop 46. European Centre for the Validation of Alternative methods. *Alternatives to Laboratory Animals,* 29, 649–668.
50. Forbes, I.I. (2000). Human airway epithelial cell lines for *in vitro* drug transport and metabolism studies. *Pharmaceutical Science and Technology Today*, 3(1), 18–27.
51. Banker, M.J., Clark, T.H., Williams, J.A. (2003). Development and validation of a 96-well equilibrium dialysis apparatus for measuring plasma protein binding. *Journal of Pharmaceutical Sciences*, 92, 967–974.
52. Jepson, G.W., Black, R.K., McCafferty, J.D., Mahle, D.A., Gearhart, J.M. (1994). A partition coefficient determination method for nonvolatile chemicals in biological tissues. *Fundamental and Applied Toxicology*, 22, 51–57.

53. Coecke, S., Ahr, H, Blaauboer, B.J., *et al.* (2006). ECVAM. metabolism: a bottleneck in *in vitro* toxicological test development. *Alternatives to Laboratory Animals*, 34, 49–84.
54. Langowski, J., Long, A. (2002). Computer systems for the prediction of xenobiotic metabolism. *Advanced Drug Delivery Reviews*. 54(3), 407–415.

9

Nanotechnology and Nanomaterial Personal Care Products: Necessary Oversight and Recommendations

*George A. Kimbrell**

International Center for Technology Assessment,
Washington, DC, USA

9.1 Introduction

Think of a woman shopping at a department store's cosmetics counter and purchasing a high-end cosmetic product, a face cream. This particular

---

*Staff Attorney, The International Center for Technology Assessment (ICTA), Washington, D.C. ICTA is a non-profit, bi-partisan organization committed to providing the public with full assessments and analyses of technological impacts on society. ICTA explores the environmental, human health, economic, ethical, social and political impacts that can result from the applications of technology or technological systems. Mr. Kimbrell works on legal developments in biotechnology, nanotechnology, and climate change technologies. He has published a number of articles and commentaries on the oversight of nanotechnology. He also drafted the first-ever legal action on the risks of nanotechnology, a petition filed with the Food and Drug Administration on behalf of a coalition of eight consumer, health, and environmental organizations in May 2006. Mr. Kimbrell joined ICTA following a clerkship with the Honorable Ronald M. Gould, United States Court of Appeals for the Ninth Circuit. He received his law degree from Lewis and Clark Law School, graduating *magna cum laude* with a Certificate in Environmental and Natural Resource Law. The views expressed herein are those of the author, and not necessarily those of ICTA or its clients. He can be reached by email at gkimbrell@icta.org.*

C. I. Betton (ed.), Global Regulatory Issues for the Cosmetics Industry Vol. 1, 117–153
© 2007 William Andrew Inc.

face cream is one of several cosmetics on the market that happens to contain a new form of manufactured material made using nanotechnology. Like the great majority of the public, the woman is completely unaware or knows next to nothing about nanotechnology, including that nanomaterials are being manufactured and inserted in consumer products. The woman applies her new face cream as directed and it washes off in the shower. From her shower drain the cosmetic product and its ingredients enter the household's waste stream, traversing the sewers beneath the woman's town and eventually out into the tributaries and waterways surrounding the town. Once in the natural environment and water cycle, the "free" nanomaterial, not fixed in any solid matrix, separates from the other ingredients in the face cream and interacts with different elements in the aquatic environment, working its way up the food chain.

Now imagine that this particular nanomaterial—known as carbon fullerene (C_{60}) or more commonly as "buckyballs" (after the architect Buckminster Fuller who designed geodesic shaped buildings similar in structure to C_{60})—was found by scientists to cause brain damage to fish and be toxic to other aquatic life, as well as be toxic to human liver cells at low levels. Would the woman feel safe placing this material on her body, or comfortable having it wash off, knowing it would enter the waste stream and the environment? Would she wonder about the applicability of existing laws and federal oversight to address any possible risks to her health and the environment from this material?

Most people tend to think of nanotechnology-based products and applications only in the future tense. If they know anything about nanotechnology at all, peoples' minds tend to conjure tiny nanorobots, mini-self-assemblers, nanodrug vectors, or something like that. Similarly, when picturing nanotechnology's risks, minds immediately conjure images of nanotechnology pioneer Eric K. Drexler's now infamous "Grey Goo" scenario,[1] or the predatory nanoswarms of fiction writer Michael Crichton's *Prey*.[2] Perhaps it is human nature to focus on those visions that fuel the imagination. Nanotechnology's reality is compelling but a bit more down to earth: the commercialization of the first nanomaterial-laced consumer products, from manufactured nanoparticles of zinc oxide and titanium dioxide used in sunscreens and cosmetics, to carbon nanotube-reinforced tennis rackets, to stain-resistant, nanomaterial-coated clothing. These products have arrived on market shelves in significant numbers and represent the crest of a product wave spanning many technologies. Early surveys highlight the already-broad scope of items with nanomaterial ingredients, including an inordinate number of nano-containing personal care products such as cosmetics and sunscreens. Most of these currently available nanomaterial personal care products fall under the broad regulatory jurisdiction in the US of the Food and Drug Administration (FDA).

This chapter first covers some nanotechnology "101" basics and the state of nanotechnology's present development and commercialization. It then discusses the known and potential human health and environmental hazards of nanomaterials. It then explains FDA's current stance on nanotechnology and nanomaterial consumer product testing and regulation and contrasts that view with that of the scientific community. Even though nanomaterials are known to have fundamentally different properties that create unique human health and environmental risks, FDA currently treats nanomaterial product ingredients no differently than bulk material product ingredients.

Next, the chapter discusses some oversight developments of nanomaterials in personal care products during the past year, including the filing of a legal petition with FDA calling on the agency to amend its regulations as applied to nanomaterial products.

Finally, the chapter concludes with some recommendations for nano-industries and government regarding nanomaterials in consumer products, the protection of public health and the environment, and the future of nanotechnologies.

9.2 What is Nanotechnology Anyway?
A New World of Tiny Technology

Nanotechnology is a powerful new platform technology for taking apart and reconstructing nature at the atomic and molecular level. It involves the manipulation of matter at the nanometer (nm) scale, one-billionth of a meter. The nanoscale is exceedingly tiny; it is the world of atoms and molecules. For illustration, a hydrogen atom is about 0.1 nm. A human DNA molecule, which carries genetic information in the cell nucleus, when uncoiled is, depending on the chromosome, about 2.5 cm long but only 2 nm in diameter. A human hair is huge by comparison, about 50,000 nm thick; the head of a pin is about 1 million nm across. A sugar molecule, which measures about 1 nm, is about as big in relation to an apple as the apple is in relation to the earth.

The simplest definition may be that of Richard Smalley, 1963 Nobel Prize Winner in Chemistry, who said that nanotechnology is "the art and science of building stuff that does stuff at the nanometer scale." In December

2006, ASTM International put forth the first formalized definition from an international standards organization, E 2456-06: "nanotechnology, n—A term referring to a wide range of technologies that measure, manipulate, or incorporate materials and/or features with at least one dimension between approximately 1 and 100 nanometers (nm). Such applications exploit the properties, distinct from bulk/macroscopic systems, of nanoscale components."[3]

The term "nanotechnology" is generally understood to encompass both nanotechnology and nanoscience. Here are a few other helpful definitions:

Nanoscience: The study of phenomena and manipulation of materials at atomic and macromolecular scales, where properties differ significantly from those at larger scale.

Nanotechnologies: The design, characterization, production and application of structures, devices, and systems by controlling shape and size at the nm scale.

Nanoscale: Having one or more dimensions of the order of 100 nm or less, or having at least one dimension that affects functional behavior at this scale.

Engineered/manufactured nanoparticle: A particle <100 nm engineered or manufactured with a specific physicochemical composition and structure to exploit properties and functions associated with its dimensions and exhibits new or enhanced size-dependent properties compared with larger particles of the same material.

Nanomaterial: particles or other manufactured substances (nanotubes, quantum dots, fullerenes, etc.) that exist at a scale of 100 nm or less or have at least one dimension that affects their functional behavior at this scale.

Nanoproduct: Any product that is composed of or that contains as an ingredient engineered or manufactured nanomaterial.[4]

As a platform technology, nanotechnology is used by many industries and is more properly referred to in the plural, nanotechnologies. The common features of "nanotechnologies" include control (the ability to put small quantities of matter where we want it), utilization (using the ability to manipulate matter at the nanoscale to "do stuff") and visualization ("seeing" where we put material).

9.2.1 Nano Means Fundamentally Different

But "nano" means more than just tiny manufacturing: Every chemical element has characteristic, defined properties: color, hardness, elasticity, conductivity, melting temperature, etc. However, if an object made of a particular substance is divided again and again until it falls out of the macro-world and into the nanoworld, these properties can change radically. Put a different way, it is well known that materials engineered or manufactured to the nanoscale exhibit different fundamental physical, biological, and chemical properties from bulk materials.[5] One reason for these fundamentally different properties is that a different realm of physics, quantum physics, governs at the nanoscale.[6] Another is that the reduction in size to the nanoscale results in an enormous increase of surface to volume ratio, giving nanoparticles a much greater surface area per unit mass compared to larger particles.[7] For example, a gram of nanoparticles has a surface area of a thousand square meters. Because growth and catalytic chemical reactions occur at the particle surface, a given mass of nanoparticles will have an increased potential for biological interaction and be much more reactive than the same mass made up of larger particles, thus enhancing intrinsic toxicity.[8] This enormous increase in surface area can change relatively inert substances into highly reactive ones. A material then can melt faster, absorb more, or simply become more explosive.

Thus, to say that a substance is "nano" does not merely mean that it is tiny, a millionth of a meter in scale; rather, the prefix is best understood to also mean that a substance has the capacity to act in fundamentally different ways. Altered properties can include color, solubility, material strength, electric conductivity, and magnetic behavior. For example, a gold wedding ring is yellow in color; but 25 nm gold nanoparticles appear red. Slightly larger gold nano spheres can appear orange or green.[9] Carbon (like graphite in pencil lead) is relatively soft; but carbon in the form of carbon nanotubes (nanoscale cylinders made of carbon atoms) is a hundred times stronger than steel. Aluminum foil will not burn if you place a lighter under it; however, aluminum nanoparticles are explosive catalysts used in rocket fuels.

9.2.2 Manufactured and Engineered Nanomaterials vs. Natural Nanoparticles

Humans and animals have been encountering naturally occurring nano-materials for millions of years, like salt nanocrystals found in ocean air or carbon nanoparticles emitted from fires. The makers of stained glass unknowingly used nanoscale metal properties for centuries to produce the beautiful colors of stained glass. However, it is only recently that scientists have developed the techniques for synthesizing and characterizing many new materials with at least one dimension on the nanoscale. To be sure, nanomaterials now in development and manufacture are different from anything that exists in nature: the very reason that nanotechnology is hyped so heavily is because it allows people to create products that do things that normal scale substances cannot. These new manufactured and engineered nanoparticles, the very building blocks of these new technologies, are patented for their novelty.[10] Accordingly, the assessment of environmental and human health risks associated with nanomaterials, which is discussed later, is largely regarding the new materials that are being so formed and generated, the increased exposure levels from engineered nanostructures now being manufactured and marketed in greater and greater quantities, and the new routes/scenarios by which human and environmental exposure can occur with the current and anticipated nanomaterial applications.

9.2.3 The Next Industrial Revolution? The Stages of Nanotechnology's Predicted Development

Before getting into the current uses of nanotechnology, it is important for perspective to step back and look with a longer lens at the topic. Nanotechnology has been touted as nothing less than the next industrial revolution: transforming and constructing a wide range of new materials, devices and technological systems in a wide number of fields including food and agriculture, electronics/computers, medicine, military applications, environmental remediation, communications, and many more. As such, nanotechnology is considered a "platform" or enabling technology.

Dr. Mike Roco of the National Nanotechnology Initiative (NNI) subdivides the predicted development of nanotechnologies into four "phases."[11] According to Dr. Roco, the first, "passive" phase of first generation products began around 2000, and include dispersed and contact manufactured nanostructures and nanoparticles such as aerosols and colloids (small agglomerations of molecules), and products incorporating manufactured nanomaterials such as nanocoatings and nanocomposites. This is the phase of the vast majority of the commercialized nanomaterials at the present. The second phase of "active nanostructures" was predicted by Dr. Roco to begin around 2005, and includes bioactive nanostructures (such as targeted drugs and biodevices) as well as physicochemically active nanostructures (such as 3D transistors, amplifiers, and adaptive structures). Further, "phases" of development may include systems of nanostructures including guided assembly and nano-robotics (*ca.* 2010) and molecular nanosystems (*ca.* 2015–2020).[12]

9.3 Nanomaterials in Consumer Products: The Future is Now

The applications of nanotechnology commercialized thus far lack the science fiction element proposed by some quarters (that people's imaginations often begin with), such as the ability to upload your brain into a super-computer (envisioned by the US National Science Foundation), the creation of molecular manufacturing plants the size of a microwave capable of producing weapons or computers (as envisioned by the Foresight Institute). While many still envision far-off in the future, nanorobots as the face of nanotechnology, the truth is that the first wave of nanotechnology is upon us now and has the much more familiar, mundane, and deceptive face: that of nano-"enhanced" consumer products. But make no mistake: nanotechnology is no longer "on the horizon." It is fast becoming a fact of daily life.

9.3.1 Measures of Nanotechnology's Maturation

One method for measuring nanotechnologies' maturation is taking a look at what has happened and what is happening. And signs of maturation abound. Most discussions begin with research and development (R&D) numbers, which are impressive, incredible even: global nanotech R&D is estimated at around $9 billion, with $1 trillion in US dollars estimated globally by 2015.[13] Investments in federally funded nanotechnology activities coordinated through the NNI were approximately $1.3 billion in 2006, and about $2 billion in annual R&D investment is currently being spent by non-federal sectors such as states, academia, and private industry. State governments spent an estimated $400 million on facilities and research aimed at the development of local nanotechnology industries in 2004. Unfortunately, only a paucity of the exuberant federal funding—4% of the NNI's FY07 budget—is earmarked for environmental health and safety (EHS) research.[14] Other non-governmental estimates put the EHS funding number as actually closer to *1%*.[15]

But much more is happening than just R&D. The term itself, "nano," has rapidly become a ubiquitous buzzword in media and society.[16] As any patent lawyer will tell you, the "gold rush" for patents on the building blocks of the platform technology continues unabated.[17] And finally nanotechnology commercialization is moving forward at lightening speed. Thousands of tons of nanomaterials are already being produced each year.[18] Consumer products containing nanomaterials have been and continue to enter the market at a steady pace. According to Lux Research's 2006 Nanotechnology Report, more than $32 billion in products incorporating nanotechnology were sold last year, more than double the previous year.[19] Lux predicts that by 2014, $2.6 trillion in manufactured products will be nanoproducts, 15% of total global manufacturing.

In the absence of mandatory nanoproduct labeling, it is very difficult to track and analyze the specific nanoproducts currently on the market. And it is becoming more difficult, as manufacturers vacillate between truthfully labeling products for the marketing purposes of buzzword hype and deciding not to label based on perceived negative reactions to nanotechnology's unknowns. However, several inventories and studies confirm that such products are available in significant numbers. In March 2006, the Project on Emerging Nanotechnologies at the Woodrow Wilson International Center for Scholars launched the first searchable inventory of nanotechnology-based consumer products, finding over 200 nano-containing products on US store shelves.[20] By the end of 2006, the Wilson Center's database numbers had nearly doubled, to over 380 products. The Wilson Center inventory was complied using only English language Internet searches and is limited to only those products self-identified. Still, the nanoproducts found include paints, coatings for eyeglasses and cars, sunscreens, medical bandages, sporting goods like tennis racket and golf balls, cosmetics, stain-resistant clothing, dietary supplements, food and food packaging, and light-emitting diodes used in computers, cell phones, and digital cameras.[21] US-based companies manufacture the majority of the products; although this may be due to the fact the database is limited to English-language products. Although some nanoproduct claims do not pass the laugh test and are clearly mislabeled for hype purposes, a far greater number of nanomaterial consumer products are not labeled at all, an untruth by omission that more than compensates for those mislabeled products.

Notably, the personal care industry stands out as perhaps the leading manufacturing sector incorporating nanomaterials into its products. The Royal Society and Royal Academy of Engineering first noted the prevalence of nanomaterial production for personal care products in its seminal 2004 report on nanotechnology.[22] Many of the nanoproducts in the Wilson Center Database are intended for human consumption, either directly or indirectly, through lotions, sunscreens, and cosmetics that are absorbed by the skin.[23] The largest database category is health and fitness products, which includes the subcategories of cosmetics, personal care, and sunscreens.

Other studies and inventories have focused specifically on nanopersonal care products. The Australian Therapeutic Goods Administration (TGA) concluded in February 2006 that there are approximately 400 nanosunscreen products containing manufactured nanoparticles of zinc oxide or titanium dioxide currently on the market in Australia.[24] In May 2006, Friends of the Earth (FoE) published a report on the incidence of nanomaterials in personal care products, detailing 116 currently available cosmetics, sunscreens and other personal care products that incorporate nanomaterials.[25] The nanomaterials found include nanoscale metal oxides such as titanium dioxide and zinc oxide, carbon spheres such as fullerenes, and nanocapsules designed to reach deep layers of the skin.[26] The product manufacturers include numerous well-known personal care companies such as Johnson & Johnson, Chanel, Estee Lauder, Revlon, L'Oreal, and others.[27] Finally, in October 2006 the Environmental Working Group (EWG) completed a survey of ingredients used in 25,000 personal care products, and found use of nanoscale materials in 256 products containing one or more of 57 different types of nanomaterial ingredients.[28]

A "spotlight" product might be helpful: nano-sunscreens. Sunscreen manufacturers found that, unlike bulk-sized amounts of the same substances, UV blockers titanium dioxide and zinc oxide became transparent (or "cosmetically clear") when manufactured at the nanoscale. The new optical properties of the nanoparticles made the clear sunscreens more marketable than those made using bulk material versions. These manufactured nanoparticle ingredients are patented for their novelty.[29]

9.4 What are the Human Health Risks of Nanotechnology and Nanomaterials in Personal Care Products?

Just as the size and chemical characteristics of engineered nanoparticles can give them exciting properties, those same new properties—tiny size, vastly increased surface area to volume ratio, and consequently potentially higher reactivity—can also create unique and unpredictable human health and environmental risks.[30] Swiss Insurance giant Swiss Re noted that "Never before have the risks and opportunities of a new technology been as closely linked as they are in nanotechnology. It is precisely those characteristics which make nanoparticles so valuable that give rise to concern regarding hazards to human beings and the environment alike."[31] The very qualities that make nanoparticles commercially desirable can also make them more harmful than their bulk material counterparts. Due to the embarrassing paucity of federal funding for EHS risk research, and the government and industry's heavy emphasis and rush to capitalize on commercialization of these materials, there is much we do not know about the affects of nanomaterials on people and the environment. To be sure, while they have the capacity to be fundamentally different, not all nanoparticle and nanomaterials are toxic or dangerous; that said, as discussed further in the coming sections of this chapter, it is crucial to understand that they are also not uniformly safe and that their safety cannot be assumed from any testing and/or profiles of their bulk material counterparts. Rather, nanomaterials present novel health and environmental risks that cannot be predicted from conventional materials. The uniqueness that excites industry, their very "nano-ness," makes them fundamentally different substances for which safety testing must be looked at anew, with nano-specific properties taken into account.

There are numerous foreseeable risks that arise from the fundamentally different nature and properties of these materials. And there is a growing body of scientific evidence finding red flags, fleshing out some of the potential dangers to humans and the environment.[32]

9.4.1 Nanotoxicity

First, as noted above, nanoparticles' exceptionally large relative surface area creates increased surface reactivity and enhanced intrinsic toxicity.[33] Nanotoxicology is an emerging field, but existing literature suggests clearly that nanoparticles have a greater risk of toxicity than larger particles, in part because as particles get smaller, their surface area to volume ratio increases and nanoparticles can behave biologically more like solutions or gases than solids. The greater a surface area to volume ratio is, the higher a substance's chemical reactivity and potentially its biological activity.[34] The greater chemical reactivity of nanoparticles can result in increased production of reactive oxygen species (ROS), including free radicals.[35] ROS production has been found in a diverse range of nanomaterials including carbon fullerenes, carbon nanotubes and nanoparticle metal oxides.[36] ROS and free radical production is one of the primary mechanisms of nanoparticle toxicity; it may result in oxidative stress, inflammation, and consequent damage to proteins, membranes and DNA.[37] Size is therefore a key factor in determining the potential toxicity of a particle. Other factors influencing toxicity include shape, chemical composition, surface structure, surface charge, aggregation and solubility.[38]

Many types of nanoparticles have proven to be toxic to human tissue and cell cultures, resulting in oxidative stress, inflammatory cytokine production, DNA mutation, and even cell death.[39] Nanoparticles such as titanium dioxide and zinc oxide used in large numbers of sunscreens, cosmetics, and personal care products have been shown to cause far greater cell damage than larger particles of the same substances. Whereas 500 nm titanium dioxide particles have only a small ability to cause DNA strand breakage, 20 nm particles of titanium dioxide are capable of causing complete destruction of super coiled DNA, even at low doses and in the absence of exposure to UV.[40] The potential for sunscreens and cosmetics containing nanomaterials to result in harm is made greater as production of ROS and free radicals increases with exposure to UV light,[41] as does related DNA damage.[42]

The first scientific evidence of exposure to nanoparticle titanium dioxide resulting in production of ROS in human brain cells was published in 2006, although it is still unknown whether the brain cells' release of ROS results in neuronal damage.[43] Titanium dioxide nanomaterials have been shown to cause oxidative stress-mediated toxicity in a range of different cell types, including skin fibroblasts, human colon cells, and rat liver cells.[44]

One of the more alarming cases found is that of carbon fullerenes (C_{60} buckyballs), used in face creams and moisturizers,[45] have been shown to be toxic to cultured human liver carcinoma cells (HepG2) at low levels of exposure.[46] Buckyballs have also been found to cause changes to the brain in fish,[47] are toxic to *Daphnia* (water fleas) and have bactericidal properties.[48]

9.4.2 Unprecedented Mobility

In addition to toxicity concerns, due to their size, nanoparticles have unprecedented mobility for a manufactured material.[49] They readily enter the human body and gain access to the blood stream via inhalation and ingestion.[50] It also appears likely that nanoparticles can penetrate the skin, although the jury is still out on the question and more research is needed. Once inside the body, nanoparticles can cross biological membranes, cells, tissues, and organs more efficiently than larger particles.[51] Once in the blood stream, nanomaterials can circulate throughout the body and can be taken up by the organs and tissues including the brain, liver, heart, kidneys, spleen, bone marrow, and nervous system.[52] In addition, unlike larger particles, nanoparticles are transported within cells and taken up by cell mitochondria and the cell nucleus, where they can interfere with cell signaling, induce major structural damage, including DNA damage.[53]

Inhaled nanoparticles are more likely than larger particles to penetrate the protective lining of the human lungs and reach the alvaeoli.[54] Animal studies have shown lung inflammation, oxidative stress, and negative impacts in other organs and the cardiovascular system following inhalation of engineered nanoparticles.

9.4.3 The Public at Large

The public at large is exposed to manufactured nanomaterials by the purchase and use of nano-containing products. The risk of exposure is clearly increased when the product is applied to the skin, like many personal care products containing nanomaterials that are used daily or even more frequently: sunscreens, toothpastes, soap, deodorants, hair shampoos and conditioners, lipsticks, face powders, antiwrinkle creams, blush and eye shadow, nail polish, and moisturizers. The nanomaterial ingredients in these products are "free" and not fixed in a solid matrix.[55] These personal care products also may be inhaled and are often ingested. Also, various food, food additives, supplements, vitamins, and food packaging products also contain nanomaterials.[56] Because nanomaterials used in foods, dietary supplements, cosmetics, color and food additives, and drugs can migrate to and through the environment,[57] their hazards can be disproportionate to their use, affecting many consumers who are not even choosing to buy those products. Finally, the public will be exposed to nanoparticles as a result of emissions from nanomaterial manufacturing processes—"nanopollution".

9.4.4 The Question of Skin Penetration

Despite the rush to commercialization and widespread consumer use, the jury is still out on the ease of skin penetration of these nanomaterials. If nanoparticles are able to penetrate the stratum corneum (outer layer of dead skin cells) and gain access to the living cells within the epidermis and the dermis, they could potentially enter the blood stream and circulate around the body with uptake by cells, tissues, and organs.[58] As the Royal Society has called for, additional research into the influence of skin condition, including sunburn, on the uptake of nanomaterials, especially in the assessment of nanomaterials found in sunscreens and cosmetics, is needed.[59] It is known that broken skin is an ineffective barrier and enables much larger particles to reach living tissue;[60] this suggests that acne, eczema, abrasions, or shaving nicks are likely to enable nanoparticle uptake. Similarly, the ability of larger particles to access the dermis through flexed skin has been demonstrated, suggesting uptake of much smaller nanoparticles is possible in some circumstances. The fact that many cosmetics and personal care products are used on blemished skin or after shaving has been largely ignored in the discussion of skin uptake. In addition, multiple products can be and are often used at the same time, creating questions about "penetration enhancing" ingredients or base carriers used in some personal care products promoting absorption of nanomaterials. *Skin Deep*, a report by the Environmental Working Group on the health risks of commercially available cosmetics and personal care products, found that more than half of all cosmetics contained ingredients that can act as penetration enhancers.[61] Thus, testing of skin uptake should be undertaken in the context of the whole products, in "real life" conditions such as flexing.[62] Size is but one factor that may influence skin uptake. Other factors that also require research include particle shape, surface characteristics, electronic charge and dose.[63] Studies investigating the skin penetration of other nanomaterials, for example titanium dioxide and zinc oxide, do not adequately investigate key variables that influence skin penetration as described above.

9.4.5 Nanomaterial Worker and Workplace Risks

Workers handling nanomaterials represent one very key class of the public the most at risk. Workers handling nanomaterials are likely to be exposed much more and at much higher levels than the general public. Workers can be exposed to nanomaterials during research, development, manufacture, packaging, handling, transport, and use of nanotech products. The numbers of workers currently exposed is unknown, although the US National Science Foundation has estimated that by 2015 there will be 2 million workers globally employed by nanotechnology industries. Current levels of existing exposure are also unknown throughout all sectors of the production chain. There is also no method currently for limiting, controlling, or even measuring human exposure to nanomaterials in the workplace. Dangers to workers could come from inhalation (the new asbestos?), access through the skin and digestive system (and the creation of free radicals that cause cell damage), or through the exposure to a new substance created through nanotechnology to which the body has no natural immunity or that triggers autoimmune disorders. Carbon nanotubes in particular have been likened to asbestos fibers and found to cause lung inflammation and granulomas in animals.[64]

9.5 What are the Environmental Risks of Nanotechnology and Nanomaterials in Personal Care Products?

As with potential direct impacts on human health, various potentially damaging environmental impacts stem from the novel nature of manufactured nanomaterials. The nanomaterials now being manufactured, marketed, and purchased are inevitably ending up in the natural environment. Entry can happen accidentally or intentionally over the course of a nanomaterials' lifecycle, during manufacturing,[65] transportation, use, recycle, or disposal. In addition, some researchers plan to introduce nanomaterials deliberately into the natural environment for environmental remediation purposes. As the many industries involved in nanotechnology expand and increase in product number and variety, both industrial and domestic nanowaste and nanopollution will logically increase in quantity as well. Once loose in nature, nanomaterials constitute a completely new class of manufactured non-biodegradable pollutants. However, despite moving at pace with nanomaterial applications and commercialization, only a few studies on the environmental impacts of engineered nanoparticles exist or are available in the public domain, leaving many potential risks dangerously untested.

Generally, nanoparticles have the ability to persist and reach places that larger particles cannot.[66] Because of their tiny size, nanoparticles move with great speed through aquifers and soils and settle more slowly larger particles. Nanoparticles originating from industrial processes, consumer products, or other sources could easily be transported by runoff or rain to water bodies. Because of their large surface area, nanoparticles provide a large and active surface for absorbing smaller contaminants, such as cadmium and organics. In the soil, nanoparticles can bond with pollutants and transport them, causing the pollutants to be absorbed by soil in larger quantities and at a faster rate. The enhanced bonding and mobility characteristics of nanoparticles create a means by which ordinarily less mobile pollutants like fertilizers or pesticides could "hitch a ride" over long distances.[67] Similarly, nanoparticles could provide a means for long-range transport of pollution in underground water, similar to colloids.[68] Further, because nanoparticles tend to be more reactive than larger particles, interactions with substances present in the soil could lead to new and possibly toxic compounds.

Even if nanoparticles are "fixed" inside a product matrix, "highly durable" nanomaterials will remain in nature long after the disposal of their host products.[69] The longevity of nanomaterials theoretically could create accumulation that could upset ecological balances, even if that particular nanomaterial is harmless to humans. One such example is manufactured nanoparticles of silver. Nanoparticles of silver are currently being used in a plethora of products, including washing machines and food packaging.[70] The same property that makes these nanoparticles attractive to manufacturers—their highly enhanced antimicrobial properties—can be highly destructive to ecosystems.[71]

Studies have also suggested that some nanomaterials bioaccumulate in microorganisms and plants. For example, scientists have found that engineered nanoparticles of aluminum oxide slowed the growth of roots in at least five species of plants: corn, cucumber, cabbage, carrot, and soybean, suggesting that nanoparticles may have adverse effects in plants as well as animals.[72] Seedlings can interact with the nanoparticles and stunt their growth.[73] Sunscreens and coatings commonly use such nanoparticles.[74]

Even simply detecting and measuring engineered nanomaterials in the environment is a new challenge created by their unique physical and chemical characteristics.[75] The methods and protocols needed are just beginning to be developed. Most technologies to measure particle size can only measure particles in micrometers, many hundreds to thousands times larger than nanoparticles.[76] Once detected, to remove them from water or air requires new filtering techniques. Nanoparticles pass through most available filters, such as those used to filter drinking water.[77] Extraction could pose a challenge because of the nanoparticles' strong adsorption properties.

9.5.1 Huge EHS Unknowns

Despite what we do know, in general, there is an alarming lack of data on environmental, health and safety issues surround nanomaterials.[78] Just a few of the more glaring gaps: methods to evaluate the toxicity of nanomaterials need to be urgently developed.[79] There are no data on human and environmental exposure levels of manufactured nanomaterials. There are no published studies relating to the absorption of carbon nanomaterials through human skin, despite the increasing incidence of fullerenes in face creams and moisturizers.[80] The length of time that nanoparticles may remain in vital organs and what dose may cause harmful effect is unknown. No research has been done on the effect on the gastrointestinal tract, even though many nanomaterial ingredients in consumer products (including food and dietary substances) will enter that area of the body.[81] Instruments to assess exposure in air and water need to be developed.[82]

9.6 FDA's Regulatory Stance on Nanotechnology and Nanomaterial Personal Care Products

The FDA defines the term nanotechnology informally on its website with reference to the definition of the NNI, as including:

1. The existence of materials or products at the atomic, molecular or macromolecular levels, where at least one dimension that affects the functional behavior of the drug/device product is in the length scale range of approximately 1–100 nm.
2. The creation and use of structures, devices and systems that have novel properties and functions because of their small size.
3. The ability to control or manipulate the product on the atomic scale.[83]

According to the agency, "nanotechnology relevant to the FDA might include research and technology development that both satisfies the NNI definition and relates to a product regulated by FDA."[84]

FDA regulates "products, not technology." [85] As discussed earlier, many consumer nano-products currently on market fall under the broad regulatory umbrella of FDA, which is charged with regulating the safety and effectiveness of most food, drugs, and cosmetics, as well as other substances such as medical devices, radiation-emitting products, animal feed, and combination products.[86] In addition to the growing number of consumer products, over 100 nanomaterial drugs and medical devices are undergoing animal or clinical trials.[87] FDA is aware of "several FDA regulated products [that] employ nanotechnology," including "cosmetic products claim[ing] to contain nanoparticles to increase the stability or modify the release of ingredients" and "nanotechnology-related claims made for certain sunscreens."[88]

FDA's jurisdiction can be divided into sections. Drugs, biologics, and medical devices require premarket approval from FDA.[89] Such approvals are rigorous and the burden of proof is on the manufacturer.[90] FDA must insure that drugs are "safe and effective."[91] Sunscreens, including nanosunscreens, are classified as human drugs because they make health claims.[92] In addition, food, drugs, nor cosmetics can be adulterated or misbranded.[93] However, in contrast to other substances, FDA has relatively limited authority over cosmetics, including potentially high-risk nanocosmetics, which does not include premarket approval.[94]

With regard to its oversight of nanomaterial products, FDA has stated that it

> believes that the existing battery of pharmacotoxicity tests is probably adequate for most nanotechnology products that we regulate. Particle size is not the issue. As new toxicological risks that derive from new materials and/or new conformations of existing materials are identified, new tests will be required.[95]

Elsewhere, FDA has concluded similarly, albeit less conclusively, that its existing regulatory requirements for products *"may be* adequate for most nanotechnology products that we will regulate."[96]

Finally, FDA has acknowledged that it has "only limited authority over some potentially high-risk products, e.g. cosmetics," and has "comparably few resources available to assess the risks of these products. . . . Few resources currently exist to assess the risks that would derive to the general population from the wide-scale deployment of nanotechnology products."[97]

9.7 Nanotoxicology: Nano-specific Testing Paradigms Are Required

FDA seemingly recognizes the fundamentally different characteristics of nanoparticles in its informal adoption of the NNI definition of nanotechnology, which includes the requirement of "the creation and use of structures, devices, and systems that have novel properties and functions because of their small size."[98] Yet FDA's existing testing methodologies are based on bulk material or larger particles, and the agency assumes that this battery of testing is "probably adequate" for testing the safety of manufactured nanoparticles.[99]

The agency's conclusion is inaccurate and is at loggerheads with the consensus view of the scientific community: "experts are overwhelmingly of the opinion that the adverse effects of nanoparticles cannot be reliably predicted or derived from the known toxicity of the bulk material."[100] For example, the European Commission's Scientific Committee on Emerging and Newly Identified Health Risks (SCENIHR) concluded: "experts are of the *unanimous* opinion that the adverse effects of nanoparticles cannot be predicted (or derived) from the known toxicity of material of macroscopic size, which obey the laws of classical physics."[101] Similarly, the UK Royal Society and the Royal Academy of Engineering emphasized: "free particles in the nanometre size range do raise health, environmental, and safety concerns and their toxicology *cannot be inferred* from that of particles of the same chemical at a larger size."[101] And finally, the British Institute for Occupational Medicine similarly concluded:

> Because of their size and the ways they are used, they [engineered nanomaterials] have specific physical-chemical properties and therefore may behave differently from their parent materials when released and interact differently with living systems. *It is accepted, therefore, that it is not possible to infer the safety of nanomaterials by using information derived from the bulk parent material.*[103]

Toxicology normally correlates health risks with the mass to which an individual is exposed, resulting in an accumulated mass as an internal dose/exposure. However, the biological activity of nanoparticles is likely to depend on physicochemical characteristics that are not routinely considered in toxicity screening studies.[104] There are many more factors affecting the toxicological potential of nanoscale materials, up to at least 16 in fact, including: size, surface area, surface charge, solubility, shape or physical dimensions, surface coatings, chemical composition, and aggregation potential—a "far cry from the two or three usually measured."[105] Unless we perform thorough investigations of all variables, we have no idea about the toxicity or safety of various products. Size is one of many factors, but is crucial: The relevance of the nanosize is that unlike larger particles, we cannot predict the toxicity of nanomaterials from the known properties of larger substances.

In short, FDA is wrong that existing tests are "probably adequate." Current testing is not totally useless, but rather that it is insufficient alone, because it does not take into account new parameters necessary. FDA's established methods of safety assessments must be significantly modified in order to address the special characteristics of engineered nanoparticles.

In fact, nanotoxicology is an emerging field in its own right, underscoring the differences of nanomaterial toxicity. In an agenda-setting 2006 article in *Nature*, 14 international nanotechnology scientists put forth nanotechnology's five "grand challenges," which included the urgent need to develop methods for assessing nanotoxicity.[106] Two recently published articles suggest new paradigms of predictive toxicology for engineered nanoparticle testing.[107] FDA should develop a basic screening framework to guide its testing, such as the tiered approach that would start with non-cellular tests to establish particle reactivity, followed by *in vitro* and *in vivo* tests for exposure pathways that are relevant to a chemical's anticipated use patterns and lifecycle.[108]

All nanomaterials' characteristics—including hazardous traits—must be learned anew by direct experimentation and cannot be inferred from existing testing completed on larger particles. This is a fundamental paradigm shift that scientists recognize and that should be similarly recognized and initiated by FDA. ICTA calls on FDA to do just this, among other things, in the legal petition discussed directly below. Until such time, nanomaterials should be considered new substances for regulatory purposes, as the UK Royal Society and Royal Academy of Engineering recommended:

> Substances made using nanotechnology should be considered new chemicals and undergo extra safety checks before they hit the market to ensure they do not pose a threat to human health. . . We recommend that chemicals produced in the form of nanoparticles and nanotubes be treated as new chemicals . . . .[109]

9.8 Nanomaterial Oversight Developments from FDA

In August 2006, FDA created an internal Nanotechnology Task Force charged with evaluating regulatory approaches with regard to the development of safe and effective nanotechnology products; and identifying and determining ways to address knowledge or policy gaps.[110] The Task Force is schedule to release a report in July 2007 with its recommendations.

In October 2006, FDA held its first-ever Public Meeting on Nanotechnology, seeking information about nanomaterial products under FDA's jurisdiction and emerging scientific issues, including safety issues.[111] The meeting was scheduled to "help FDA further its understanding of developments in nanotechnology materials that pertain to FDA-regulated products."[112] FDA received over 10,000 public comments expressing concern that the agency is so far behind the industry and over the potential human health and environmental impacts of nanomaterials.

There were several catalysts to this oversight activity. First and most obviously was the explosion of nanomaterial consumer products under FDA's jurisdiction.[113] Second was the German "Magic Nano" incident: in April 2006 German authorities recalled a purported nanomaterial bathroom cleaner aerosol spray after over 110 users had difficulty breathing and 6 people were hospitalized with lung edema. At first, it was feared that nanoparticles in the spray had caused the health complaints.[114] The product was "nano" because the coating that it left on sprayed surfaces was only a few nanometers thin; however, after a two and a half month investigation by the German authorities, it was found that the product did not contain manufactured nanoparticles. Instead, a liquid ingredient, atomized to give ultrafine droplets by the propellant gas, was to blame. The lessons from Magic Nano for US regulators, it would seem, are simple: the lack of adequate standardization and labeling allows for fraudulent or misleading marketing of products which can confuse the public and significantly delay or detour urgent health investigations, should the product cause a health or environmental hazard. In any event, the scare no doubt affected FDA's flurry of activity.

Third, in May 2006, the International Center for Technology Assessment (ICTA) and a coalition of consumer, health, and environmental groups[115] filed a formal legal petition with the FDA, calling on the agency to address the human health and environmental risks of nanomaterials in consumer products.[116] The petition is the first US legal action filed on the potential human health and environmental risks of nanotechnology.[117] The petition documents the existing body of scientific evidence studying nanomaterial risks stemming from their unpredictable toxicity and seemingly unlimited mobility. It requests the FDA to:

- issue a formal Commissioner opinion[118] on engineered nanoparticles in light of evidence;

- amend its regulations to include nanotechnology definitions necessary for proper regulation;
- enact comprehensive nanoproduct regulations, including treating nanomaterials as new substances, using nanomaterial-specific toxicity testing paradigms, and requiring nanoproduct labeling;
- comply with the National Environmental Policy Act (NEPA) by assessing the human health and environmental impacts of its nano-related actions. These regulations need to be retroactive in order to cover existing nanoproducts.

The petition's second half focuses specifically on nano-sunscreens.[119] Sunscreens are classified by FDA as human drugs, unlike many other personal care products, and consequently can be subject to more rigorous FDA regulation, including the requirement of premarket "new drug product" applications supporting the drug's safety and efficacy.[120] The commercial allure of nano-sunscreens is that they appear transparent because of the nanoparticles' fundamentally different properties. The engineered nanoparticles are also patented for their profitable novelty. Yet in the agency's first and only word on sunscreens, a 1999 regulation, FDA considered engineered nanoparticle ingredients in these sunscreens a mere reduction in size and not a new drug ingredient, permitting sunscreen manufacturers to sell nanosunscreens based on the safety assessment of bulk material sunscreens.[121] The petition asks FDA to reconsider its 1999 equivalency stance on nanoparticle sunscreen ingredients and for the agency to instead classify nanosunscreens as new drug products which require premarket review of health and safety evidence.[122] Nanoparticle ingredients in sunscreens have raised red flags for scientists because it is unknown how easily they can penetrate the skin and circulate throughout the body,[123] and studies have shown them to be photoactive in some cases, producing free radicals and causing DNA damage to human skin cells.[124] Because nano-sunscreens are currently sold without such premarket testing or review by FDA, the petition asks FDA to declare that those products an imminent hazard to public health and to request that manufacturers cease production until FDA nanomaterial product regulations are developed and implemented.[125]

Finally, while not a catalyst to FDA oversight development, an October 2006 Woodrow Wilson International Center for Scholars Report of the Project on Emerging Nanotechnologies is worth noting. The report, written by a former FDA Deputy Commissioner for Policy Michael Taylor was critical of FDA's preparedness and the adequacy of FDA's existing oversight framework, concluding that FDA was not "nano ready" and needed to take immediate steps to address the first wave of nanomaterial consumer products.[126] Taylor highlighted the regulatory gaps that exist in FDA's authority as applied to nanomaterials, particularly in the areas cosmetics and dietary supplements, and recommended that FDA request cosmetic companies submit safety substantiation data.[127] Taylor also recommended that FDA establish criteria for nanomaterials, including "new for legal and regulatory purposes" and "new for safety evaluation purposes."[128]

FDA procedurally responded to ICTA's petition in November 2006, pointing to its Task Force and Public Meeting as good faith efforts that it was working on the difficult and complex issue. These basic activities are a good start, albeit overdue. Unfortunately, while FDA is still conducting basic research to substantiate the safety of nanomaterial ingredients, forming preliminary task forces, and holding meetings, the use of these nanomaterials in products continues to expand.

9.9 Conclusions and Recommendations for Government and Industry

The first wave of nanomaterial products has arrived on market shelves. But engineered nanoparticles have fundamentally different properties from their bulk material counterparts—properties that also may create unique human health and environmental risks—which necessitate new health and safety testing paradigms. In the midst of this rush to increase nanoproduct application R&D, be granted nanopatents, and commercialize products as quickly as possible, serious questions regarding ethics, human and environmental heath and safety, socio-economic disruption and democracy have been ignored.[129]

Nano-personal care products present perhaps the most immediate cause for concern, given their prevalence on markets, their use of "free" rather than "fixed" nanoparticles, and their repeated, intimate use by consumers. Crucial safety questions based upon nanoparticles' toxicity and mobility characteristics as well as the existing data are still unanswered.

In order to fulfill its statutory mandates, FDA should gather input from all interested parties and begin the process of amending its regulatory framework to account for the fundamental differences and associated hazards of nanomaterials. The ICTA legal petition provides the agency both a blueprint for its possible regulatory amendments and a legal impetus to take that action. FDA now must use its inherent statutory authority to require health and safety information from the best sources: the manufacturers of these nanomaterial products.

A cosmetics industry umbrella group has called our legal petition "reckless."[130] My response is to ask what is logically reckless—demanding a regulatory agency to do its job and comply with the law on behalf of the public, calling on it to protect human health and the environment—or knowingly bringing fundamentally different and unpredictable materials into the homes and the environment without adequate oversight or independent safety review? Rather than fight oversight, nanotechnology industries should support adequate oversight—a regulatory framework that protects workers, the general public and the environment from the impacts of nanomaterials. Such a framework would include: (1) much more rigorous EHS research, (2) nano-specific toxicity testing, (3) premarket precautionary regulation, (4) transparency and informed consumer choice, and (5) oversight of nanomaterials throughout their lifecycle. Each is discussed below.

9.9.1 Support Much More Vigorous EHS Research

Inadequate data on potential environmental and human health risks, compounds the challenge of analyzing the adequacy of existing laws to nanomaterials and judging their limitations and gaps. This dearth of data stems in large part by the inadequate federal funding provided for EHS study of nanomaterials to date.[131] Publicly available, peer-reviewed, and independent study of human health and environmental impacts is urgently needed to protect public heath and the environment and provide the bases of adequate regulatory oversight of nanomaterials. Knowledge must be effectively translated into meaningful regulatory standards that reflect both current scientific understanding and critical areas of uncertainty. Crucially, the lack of data or evidence of specific harm should not be a proxy for reasonable certainty of safety. Finally, effective risk research would also address regulatory and oversight needs in addition to new scientific knowledge.[132]

9.9.2 Acknowledge the Unknowns and Fundamental Differences of Nanomaterials and Act Accordingly

While nanoparticles may consist of molecules that are regulated under existing statutes in larger forms, nanomaterials can behave very differently due to their "nano-ness" (small size, negligible mass, and higher reactivity). Accordingly, applying conventional methods alone to identify, monitor, measure, and control nanoparticles is inappropriate and insufficient. New technologies and adapted protocols for measuring, monitoring, and controlling nanomaterials are required.

Environmental policy and regulation currently relies on well known and understood chemical and physical properties, including solubility, reactivity, toxicity, and mass. Almost all environmental release restrictions and risk assessment measurements in existing environmental law are premised on a direct relationship between volume or mass and exposure. But adequate regulatory oversight and toxicity testing of nanomaterials necessitates an entirely new set of analysis factors, including, but not limited to, particle size, surface area, shape, composition, conductivity, and reactivity. Urgent study is needed to further flesh out which parameters are crucial for nanotoxicity measurements. FDA should not be passively waiting for more detailed research; rather, it should act quickly to review existing science and work with industry and public interest stakeholders to lead and accelerate the development of appropriate test protocols relevant to new applications as they are being developed.

Nanotechnology industries should acknowledge these scientific unknowns and fundamental differences and act accordingly. Industry should not disingenuously claim that current testing methods are "sufficiently robust," as it has,[133] while simultaneously admitting, as it must, that a whole new range of up to 16 new possible toxicity factors must be tested for nanomaterials.[134] This is misleading semantics. Any claims like those made for biotechnology products—of "substantial equivalence"—are belied by the universal view of the scientific community that these substances have new, novel, and fundamentally different properties.

9.9.3 Prepare to Meet the EU Standards: The Burden of Proof Is on Industry

Given that nanomaterials can pose structure-dependent health hazards unique and unpredictable from their larger bulk material counterparts, to be adequate regulation must include rigorous, accurate, and comprehensive premarket safety assessment. The precautionary principle[135] is being enacted regarding chemicals generally (including nanomaterials) by the EU under its new REACH (Registration, Evaluation, and Authorization of Chemicals) Regulation, scheduled to take effect in June 2007.[136] REACH shifts the burden of proof to the manufacturer or industry to provide information, assess risk, and provide reasonable assurances of safety prior to marketing and use, rather than placing the burden on regulators to prove harm.[137] Whether or not one disagrees with the application of the precautionary principle for existing substances, its application is even more logical and necessary when dealing with new technological advances, like those stemming from nanotechnologies, where long-term health and environmental impacts have not been adequately studied and are more unpredictable. Make no mistake, companies should prepare to generate and provide the safety data for their products to meet the EU's regulatory approach, as REACH represents a growing trend world-wide shifting the burden of proof safety from regulators to industry and away from the US's antiquated system.

Rather than fearing overregulation, industry should fear underregulation. As seen by the German Magic Nano incident, lack of adequate oversight places all nano-industries at the mercy of the bad actors, those whose products actually contain nanomaterials but have not done adequate safety testing and/or those that only purport to have nanomaterial ingredients in their products for marketing purposes. No one in government or industry sought to verify the Magic Nano claims or require safety data. Now the case is almost invariably brought up when nanotechnology is discussed. The Los Angeles Times referred to it as "a case that highlights the murky definitions and poorly understood risks in one of the fastest-growing segments of science and technology."[138] Failure to provide transparency and industry-wide standards allows for unanticipated adverse effects to cause possibly unwarranted public confidence backlashes for the whole industry. Adequate regulatory oversight would inform consumers, standardize product ingredients, and eliminate the possibility of said unwarranted nanoproduct safety recall and the accompanying public perception backlash. US nano-industries thus should be in favor of premarket testing, review, and approval and mandatory standards. The greatest risk for nano-industries is not overregulation but rather underregulation combined with reckless technology management.

9.9.4 Transparency

Industry should volunteer to FDA all information on nanomaterial ingredients in its products. The fundamental public rights to know are compromised by the lack of transparency regarding the use of nanomaterials in consumer products and the limited disclosure of information regarding safety. The public's right to know includes the right to be informed, to make educated choices, and the right to safety. Polls show that the vast majority of the public is not well informed about the presence of nano-materials in consumer products: nearly 70% have little or no knowledge about nanotechnology.[139] Neither the public, nor NGOs, nor the agencies charged with oversight responsibilities currently have a reliable means to identify products that contain manufactured nanomaterials, nor can they assess the safety of those that are identifiable. In many cases manufac-turers have not disclosed publicly health hazard and testing information; nor are they reliably labeling their product ingredients. As a result, the public has no way to make informed choices about nanomaterial products. At a minimum, consumers need transparency so they can make informed decisions about the products they buy. In addition to that, however, nano-product and ingredient labeling is necessary in order to provide causation and traceability for unanticipated health or environmental effects; create an inventory of those effects; and analyze the cumulative exposure or release effects of multiple nanoproducts.

Transparency is also needed vis-a-vis safety testing of nanomaterials. Independent, peer-reviewed testing and public dissemination of results of that testing are necessary to provide a comprehensive look at the risks and benefits of nanotechnology and therefore enable informed public discussion. While some manufacturers are no doubt conducting and have conducted rigorous safety testing, that research on safety and efficacy is being claimed as Confidential Business Information (CBI) in many cases. FDA must coordinate with other regulating agencies and develop data protocols that would protect the public interest in addition to that of private technology developers. CBI should not be permitted to include health and safety data so vital to public review and assessment of nanomaterial ingredients.

Industry officials often respond to questions of confidential safety infor-mation by asking the rhetoric question "why would anyone make an unsafe product?" First, there is a long history of knowingly making unsafe products: cigarettes and asbestos come to mind. Second, the polls clearly show that people do not trust industry with the job of protecting their health, they trust regulators. And rightly so: a family's health and safety, not to mention the well-being of the environment, should not rest on a corporate cost–benefit analysis; rather it should be entrusted to those bodies statutorily charged with protecting the public health and safety.

9.9.5 Repeating Past Mistakes: Running From Your Product's Label Is Not a Business Plan

The industry's response to negative nano press or from the hint of more rigorous regulatory oversight, appears to be to have less, rather than more, transparency. There are a number of products containing (or purporting to contain) silver nanoparticles: food storage, washing machine, refrigerator lining, shoe lining, air filters and fresheners, drywall, paint, medical coatings, and wide range of other products.[140] Silver can be highly toxic to aquatic organisms such as plankton and has the potential to bioaccumulate in some aquatic species.[141] These nanosilver particles are inserted for their antimicrobial properties; however, the same nano-enhanced properties can harm microorganisms and ecosystems, leading to questions about their environmental impacts once released in the waste stream, and calls for their regulation as pesticides.[142]

In November 2006, EPA announced that it was planning to regulate one nanosilver consumer product, Samsung's SilverCare Washing Machine, as a pesticide under the Federal Insecticide, Fungicide, and Rodenticide Act (FIFRA), because of the antimicrobial effects that products' releasing of "silver ions" during its wash cycle. EPA has premarket approval and testing authority over pesticides under FIFRA: A pesticide must be registered with the EPA before it can be distributed or sold.[143] In response to EPA's proposed action regarding nanosilver, at least one nanosilver product manufacturer, Sharper Image, removed its labeling: Sharper Image's FresherLonger™ Miracle Food Storage Containers were previously marketed as "infused with antibacterial silver nanoparticles" that were "25 nm in diameter"; however, after EPA's announcement Sharper Image stripped its website of all its print and online advertising, which now no longer makes claims to either nanosilver ingredients or biocide activity.[144]

Sharper Image's decision to seek less accountability and transparency for its nanomaterial product at the first hint of oversight does not inspire confidence that manufacturers of nanomaterial products will strive to give the public the ability to make informed choices or that the nano industries will learn the lessons of past technology failures.

9.9.6 Lifecycle

Because of their presence in all environmental media, nanomaterials affect every area of environmental concern. Environmental impacts can occur at any stage of a nanomaterials' lifecycle—R&D, manufacturing, transportation, product use, recycling, disposal, or some time after disposal—and a nanomaterial lifecycle framework helps assess how various statutory frameworks apply and where regulatory gaps exist.[145] To address all possible exposures and environmental impacts adequately, a nanomaterial's complete lifecycle must be considered.

9.9.7 See the Big Picture: The Question Is Not If, But When

Finally, adequate regulatory oversight of nanomaterials needs a clear perspective on the claims for nanotechnology. As discussed above, today's nanomaterial products represent the "first phase" or stage of nanotechnology and further phases of predicted development promise much more complicated structures. The nanoproducts currently in the market are just the tip of the iceberg. In fact, the hype and promise (and it's always difficult to separate the two) promise "nothing less than complete control over the physical structure of matter—the same kind of control over molecular and structural makeup of physical objects that a word processor provides over the form and content of a text."[146] Going forward, nanotechnology will be discussed in the context of convergence and integration of nanotechnology, biotechnology, information technology, and cognitive technology, producing more novel nanoscale materials. Thus, even if only a small portion of nanotechnology's predicted promise comes to pass, as a long-term solution it is obvious that current laws are not equipped to regulate such fundamentally different products and processes. Traditional regulatory frameworks, benchmarks, and distinctions will be less—not more—useful as applied to nanotechnology's processes and applications over time. A new nano-specific law will be needed; it is only a matter of when.[147]

9.9.8 Learn From The Past

This is not the first "wonder" material or technology that the world has seen. Throughout history substances once thought "miraculous" turned out to be deadly or harmful to the environment. Asbestos was once considered an ideal flameproof material for clothing, buildings, and other goods and was said to save lives; now it kills 10,000 people annually. For over 50 years chlorofluorocarbons (CFCs) were used as a refrigerants and insulators in innumerable household appliances, consumer products and industrial applications; scientists today know CFCs trigger ozone destruction, leading to less protection from the sun's UVB rays and increasing the risk of skin cancer, and eventually leading to international and national bans on their release. As illustrated by asbestos, CFCs, DDT, leaded gasoline,

PCBs, mercury, and numerous other former "wonder" substances and technologies, some nanomaterial will undoubtedly have significant, unforeseen, and unintended negative consequences on human health and the environment. Will policymakers and regulators wait until another such catastrophe occurs? Or will they adapt laws and regulations preemptively, in an attempt to avoid such a disaster?

As for industry, it has demonstrated its ability to innovate when government regulations require changes to standard operating procedures with strong regulatory mandates, deadlines, and enforcement tools. For example, the global ban on ozone-depleting CFCs brought the shift to safer alternatives, as did restrictions on volatile organic compounds in consumer products. If it values the future of this technology, it must stop the current race to market, focus on risk research, and work with regulators to create a premarket regulatory environment that protects consumers and the environment as well as creates stability for safe nanomaterial products.

References

1. Eric Drexler, *Engines of Creation*, Anchor Books (1986).
2. Michael Crichton, Prey (2003).
3. ASTM International, E2456-06 Standard Terminology Relating to Nano-technology, Subcommittee E56.01 on Terminology and Nomenclature, approved December 8, 2006.
4. ICTA Legal Petition on Nanotechnology to FDA, available at www.icta.org
5. National Nanotechnology Initiative, *What is Nanotechnology?* Available at http://www.nano.gov/html/facts/whatIsNano.html
6. Nanotechnology Now, Nanotechnology Basics, available at http://www.nanotech-now.com/basics.htm
7. Nel, A. *et al.* (2006), "Toxic Potential of Materials at the Nanolevel," *Science* 311, 622–627 (2006).
8. European Commission's Scientific Committee on Emerging and Newly Identified Health Risks (SCENIHR), "Opinion On the Appropriateness of Existing Methodologies to Assess the Potential Risks Associated with Engineered and Adventitious Products of Nanotechnologies, Adopted September 28–29, 2005; Warheit, D.D. (2004). Nanoparticles: Health impacts?" *Materials Today* 7, 32–35 (2004).
9. See https://www.ee.washington.edu/research/mems/group/misc/nyt-nano1.pdf
10. See generally US Patent and Trademark Office, Classification 977, Nano-technology.
11. See generally Mihail C. Roco, National Science Foundation and National Nanotechnology Initiative, "Governance of Nanotechnology for Human Development," *Presentation, Science and Technology for Human Future* (April 28, 2006); M.C. Roco, "Nanotechnology's Future," *Scientific American* (July 24, 2006).

12. *Ibid.*
13. See, e.g., Lux Research, *The Nanotech Report*, 4th Edition (2006), available at http://luxresearchinc.com/TNR4_TOC.pdf
14. See, e.g., International Center for Technology Assessment, *Congressional Letter on NNI 2006 Budget*, available at http://www.icta.org/doc/nano%20 approp%20letter_Feb_2006.pdf
15. Woodrow Wilson International Center for Scholars, Project on Emerging Nanotechnologies, Press Release, *Nanotechnology Development Suffers from Lack of Risk Research Plan, Inadequate Funding and Leadership*, September 21, 2006, available at http://www.wilsoncenter.org/index. cfm?topic_id=166192&fuseaction=topics.item&news_id=201894
16. See, e.g., Lux Research, 2006, available at http://luxresearchinc.com/ TNR4_TOC.pdf
17. See, e.g., Charles Choi, "NanoWorld: Nano Patents in Conflict," *Washington Times*, April 25, 2005.
18. See, e.g., The Royal Society and the Royal Academy of Engineering, *Nanoscience and Nanotechnologies: Opportunities and Uncertainties*, London, July 2004, pp. 26–27 and Table 4.1, available at http://www.nanotec.org. uk/finalReport.htm (hereafter Royal Society Report).
19. See, e.g., Lux Research, 2006, available at http://luxresearchinc.com/TNR4_ TOC.pdf
20. The Woodrow Wilson International Center for Scholars, Project on Emerging Nanotechnologies, *Nanotechnology Consumer Products Inventory*, available at http://www.nanotechproject.org/consumerproducts
21. *Ibid*
22. *Nanoscience and Nanotechnologies: Opportunities and Uncertainties,* The Royal Society and the Royal Academy of Engineering, London, July 2004 at Table 4.1.
23. The Project on Emerging Nanotechnologies at the Woodrow Wilson International Center for Scholars, *The Nanotechnology Consumer Products Inventory Analysis,* at 5, available at http://www.nanotechproject.org/index. php?id=44
24. Australian Government, Therapeutic Goods Administration, *Safety of Sunscreens Containing Nanoparticles of Zinc Oxide or Titanium Dioxide*, February 2006.
25. Friends of the Earth, *Nanomaterials, Sunscreens and Cosmetics: Small Ingredients, Big Risks* (May 2006), available at http://www.foe.org/camps/ comm/nanotech/nanocosmetics.pdf
26. *Ibid*
27. *Ibid*
28. Environmental Working Group, News Release, *Hundreds of Personal Care Products Contain Poorly Studied Nano-Materials*, October 10, 2006, http:// ewg.org/issues/cosmetics/20061010/index.php
29. For more on nano-sunscreens, see ICTA Legal Petition on Nanotechnology to FDA, available at www.icta.org
30. See, e.g., Andre Nel *et al.*, *Toxic Potential of Materials at the Nanolevel*, *Science*, 311, 622–627, 622, 623 and Figure 1 (2006); see generally Florini *et al.*, "Nanotechnology: Getting It Right the First Time," Nanotechnology Law & Business. 38, 41–43 (2006), at 3.

31. Swiss Re, *Nanotechnology-Small Matter, Many Unknowns* (2004), at 17.
32. See, e.g., Andrew Maynard, "Nanotechnology: The Next Big Thing, or Much Ado about Nothing?" *Annals of Occupational Hygiene* (September 2006), at 4–7.
33. See, e.g., Andre Nel et al., *Toxic Potential of Materials at the Nanolevel*, Science 311, 622–623 (2006).
34. See, e.g., Tran et al., *A Scoping Study to Identify Hazard Data Needs For Addressing The Risks Presented By Nanoparticles and Nanotubes*, Institute of Occupational Medicine Research Report (December 2005), at 21–23.
35. See, e.g., Andre Nel et al., *Toxic Potential of Materials at the Nanolevel*, Science, 311, 622 (2006).
36. G. Oberdörster et al., "Nanotoxicology: An Emerging Discipline from Studies of Ultrafine Particles," *Environmental Health Perspectives*, 113, 823–839 (2005), at 7.
37. See, e.g., Andre Nel et al., "Toxic Potential of Materials at the Nanolevel," *Science* 311, 622 (2006).
38. *Ibid.*
39. See, e.g., Friends of the Earth, *Nanomaterials, Sunscreens and Cosmetics: Small Ingredients, Big Risks* (May 2006).
40. K. Donalson et al., "Free Radical Activity Associated with the Surface of Particles: A Unifying Factor in Determining Biological Activity?" *Toxicology Letters* 88, 293–298 (1996).
41. G. Oberdörster et al., "Nanotoxicology: An Emerging discipline From Studies of Ultrafine Particles," *Environmental Health Perspectives* **113**, 823–839 (2005), at 7.
42. R. Dunford et al., "Chemical Oxidation and DNA Damage Catalysed by Inorganic Sunscreen Ingredients," *FEBS Letters*, 418, 87–90 (1997).
43. T. Long et al., "Titanium Dioxide (P25) Produces Reactive Oxygen Species in Immortalized Brain Microglia (BV2): Implications for Nanoparticle Neurotoxicity," *Environmental Science and Technology*, 40(14), 4346–4352 (2006).
44. *Ibid.*
45. Bethany Halford, "Fullerene For The Face: Cosmetics Containing C60 Nanoparticles are Entering the Market, Even if Their Safety is Unclear," *Chemical and Engineering News*, March 27, 2006, available at http://pubs.acs.org/cen/science/84/8413sci3.html
46. V. Colvin et al., "The Differential Cytotoxicity of Water-soluble Fullerenes," *Nanoletters*, 4, 1881–1887 (2004).
47. E. Oberdörster, "Manufactured Nanomaterials (Fullerenes, C60) Induce Oxidative Stress in the Brain of Juvenile Largemouth Bass," *Environmental Health Perspectives*, 112, 1058–1062 (2004).
48. See generally G. Oberdörster et al., "Nanotoxicology: An Emerging Discipline from Studies of Ultrafine Particles," *Environmental Health Perspectives*, 113(7), 823–839 (2005).
49. Hoet et al., "Nanoparticles-Known and Unknown Health Risks, *Journal of Nanobiotechnology*, 12 (December 2004), at 2; Swiss Re, *Nanotechnology-Small Matter, Many Unknowns* (2004), at 7.
50. See, e.g., Andrew Maynard, *Nanotechnology: A Research Strategy for Addressing Risk*, Woodrow Wilson International Center for Scholars, Project on Emerging Nanotechnologies, (July 2006), at 10.

51. See, e.g., Holsapple *et al.*, "Research Strategies for Safety Evaluation of Nanomaterials, Part II: Toxicological and Safety Evaluation of Nanomaterials, Current Challenges and Data Needs," *Toxicological Sciences*, 88, 12 (2005).

52. See, e.g., Oberdorster *et al.*, "Nanotoxicology: An Emerging Discipline from Studies of Ultrafine Particles," *Environmental Health Perspectives* 113, 823–839 (2005).

53. See, e.g., Andrew Maynard, *Nanotechnology: A Research Strategy for Addressing Risk,* Woodrow Wilson International Center for Scholars, Project on Emerging Nanotechnologies (July 2006), at 11.

54. G. Oberdorster *et al.*, "Deposition Clearance and Effects in the Lung," *Journal of Aerosol Medicine*, 5(3), 179–187 (1992).

55. "Fixed" nanomaterials are immobilized in a solid matrix (e.g., tennis rackets reinforced with carbon nanotubes). On the other hand, "free" nanoparticles are suspended in the liquid or cream. Free particles are more easily dispersed and more quickly spread around. Free particles are also a form more conducive to being absorbed by organisms. These types of particles make up the largest percentage of the known nanomaterial consumer product market, including cosmetics and sunscreens.

56. See, e.g., The Woodrow Wilson International Center for Scholars, Project on Emerging Nanotechnologies, *Nanotechnology Consumer Products Inventory.*

57. See, e.g., C.G. Daughton *et al.*, "Pharmaceuticals and Personal Care Products in the Environment: Agents of Subtle Change?" *Environmental Health Perspectives*, 107(Suppl 6) (December 1999).

58. See, e.g., Oberdorster *et al.*, "Nanotoxicology: An Emerging Discipline from Studies of Ultrafine Particles," *Environmental Health Perspectives*, 113, 823–839 (2005).

59. See, e.g., The Royal Society and the Royal Academy of Engineering, *Nanoscience and Nanotechnologies: Opportunities and Uncertainties*, London (2004).

60. See, e.g., Oberdorster *et al.*, "Nanotoxicology: An Emerging Discipline from Studies of Ultrafine Particles," *Environmental Health Perspectives*, 113, 823–839 (2005).

61. Environmental Working Group, Skin Deep, available at http://www.ewg.org/issues/cosmetics/FDA_warning/index.php

62. N. Monteiro-Riviere *et al.*, "Penetration of Intact Skin by Quantum Dots with Diverse Physicochemical Properties," *Toxicological Sciences*, 91(1), 159–165 (2006); S. Tinkle *et al.*, "Skin As a Route of Exposure and Sensitisation in Chronic Beryllium Disease", *Environmental Health Perspectives*, 111, 1202–1208 (2003); J. Lademann *et al.*, "Penetration Profile of Microspheres in Follicular Targeting of Terminal Hair Follicles," *Journal of Investigative Dermatology*, 132, 168–176 (2004); Oberdorster *et al.*, "Nanotoxicology: An Emerging Discipline from Studies of Ultrafine Particles," *Environmental Health Perspectives*, 113, 823–839 (2005).

63. See, e.g., Friends of the Earth, *Nanomaterials, Sunscreens and Cosmetics: Small Ingredients, Big Risks* (May 2006).

64. Andrew Maynard, *Nanotechnology: A Research Strategy for Addressing Risk,* Woodrow Wilson International Center for Scholars, Project on Emerging Nanotechnologies, (July 2006), at 8.

65. Nanomaterial manufacturing can be done by either the "top-down" or the "bottom-up" method. Top-down manufacturing is the grinding or breaking down of a substance to the nanoscale while bottom-up involves building materials through chemical synthesis including self-assembly. See generally Royal Society Report at 25 and Table 4.1.

66. See, e.g., External Review Draft, *Nanotechnology White Paper* (hereafter EPA White Paper), at 36–37, prepared for the US Environmental Protection Agency by members of the Nanotechnology Workgroup, a group of EPA's Science Policy Council. Science Policy Council, US Environmental Protection Agency, Washington, DC 20460, December 2, 2005, available at http://www.epa.gov/osa/nanotech.htm

67. *Ibid.* at 37, 40.

68. Colvin, V., *Responsible Nanotechnology: Looking Beyond the Good News*, EurekAlert! Nanotechnology in Context: November 2002, available at http://www.eurekalert.org/context.php?context=nano&show=essays&essaydate=1102

69. Andrew Maynard, *Nanotechnology: A Research Strategy for Addressing Risk,* Woodrow Wilson International Center for Scholars, Project on Emerging Nanotechnologies, (July 2006) at 12.

70. The Woodrow Wilson International Center for Scholars, Project on Emerging Nanotechnologies, *Nanotechnology Consumer Products Inventory.*

71. Letter from Ken Kirk, Executive Director, National Association of Clean Water Agencies, to Stephen Johnson, Administrator, Environmental Protection Agency (February 14, 2006) (on file with author); Letter from Chuck Weir, Chair, Tri-TAC, to James Jones, Director, Office of Pesticide Programs, Environmental Protection Agency (January 27, 2006) (on file with author). Pat Phibbs, *Pesticides: Examining Use of Nanoscale Silver in Washing Machines as Possible Pesticide*, Daily Environment Report, May 15, 2006, at A-5–A-6.

72. D. Watts, "Particle Surface Characteristics May Play an Important Role in Phytotoxicity of Alumina Nanoparticles," *Toxicology Letters*, 158, 122–132 (2005); "Study Show Nanoparticles Could Damage Plant Life," *Science Daily* (November 22, 2005), available at http://www.sciencedaily.com/releases/2005/11/051122210910.htm

73. *Ibid.*

74. The Project on Emerging Nanotechnologies at the Woodrow Wilson International Center for Scholars, *The Nanotechnology Consumer Products Inventory Analysis.*

75. See, e.g., The Royal Society and the Royal Academy of Engineering, *Nanoscience and Nanotechnologies: Opportunities and Uncertainties*, London (2004), at 42; External Review Draft, *Nanotechnology White Paper* (hereafter EPA White Paper), at 36–37, prepared for the US Environmental Protection Agency by members of the Nanotechnology Workgroup, a group of EPA's Science Policy Council. Science Policy Council, US Environmental Protection Agency, Washington, DC 20460 (December 2, 2005), at 42.

76. *Ibid.*

77. NRDC *et al.*, Comments to EPA, Re: EPA Proposal to regulate nanomaterials through a voluntary pilot program, Docket ID: OPPT-2004-0122, July 5, 2005, at 7.

78. Maynard *et al.*, "Safe Handling of Nanotechnology," *Nature* (2006). Andrew Maynard, *Nanotechnology: A Research Strategy for Addressing Risk,* Woodrow Wilson International Center for Scholars, Project on Emerging Nanotechnologies, (July 2006), at 13.
79. Andrew Maynard, *Nanotechnology: A Research Strategy for Addressing Risk,* Woodrow Wilson International Center for Scholars, Project on Emerging Nanotechnologies, (July 2006), at 28–29.
80. N. Monterior-Riviere *et al.*, "Challenges for Assessing Carbon Nanomaterial Toxicity to the Skin," *Carbon,* 44, 1070–1078 (2006).
81. Jennifer Kuzma, *Nanotechnology in Agriculture and Food Production,* Woodrow Wilson International Center for Scholars, Project on Emerging Nanotechnologies, 12 (September 2006).
82. American Bar Association, Section on Environment, Energy, and Resources, ABA SEER CAA *Nanotechnology Briefing Paper* (June 2006), at 12–14; American Bar Association, Section of Environment, Energy, and Resources, *Nanotechnology Briefing Paper Clean Water Act* (June 2006), at 4.
83. FDA, FDA and Nanotechnology Products, Frequently Asked Questions, at http://www.fda.gov/nanotechnology/faqs.html; FDA, Nanotechnology, available at http://www.fda.gov/nanotechnology/
84. *Ibid.*
85. FDA, *FDA Regulation of Nanotechnology Products,* available at http://www.fda.gov/nanotechnology/regulation.html
86. See generally The Federal Food, Drug, and Cosmetic Act (FFDCA), 21 U.S.C., Chapter 9 et seq.
87. Carrie Dahlberg, *Nanotech's Tiny Revolution Raises Caution,* Sacramento Bee (August 19, 2006).
88. FDA, Nanotechnology Products, Frequently Asked Questions, available at http://www.fda.gov/nanotechnology/faqs.html
89. See, e.g., 21 U.S.C. §§ 321(p), 355(a), 360c(a)(1)(C).
90. *Ibid.*; 21 C.F.R. Parts 200 through 499.
91. 21 U.S.C. § 393(b)(2)(B).
92. See, e.g., 58 Fed. Reg. 28195; 21 U.S.C. § 321(g)(1).
93. See, e.g., 21 U.S.C. §§ 331(a), 343(a), 352(a), 362(a).
94. FDA, Center for Food Safety and Applied Nutrition, FDA Authority Over Cosmetics (2006), available at http://www.cfsan.fda.gov/ dms/cos-206.html; 21 C.F.R. pts. 361–363, §740.10(a); 21 C.F.R. §§ 7.40–7.59.
95. FDA, Regulation of Nanotechnology Products, available at http://www.fda.gov/nanotechnology/regulation.html
96. FDA, FDA and Nanotechnology Products, *Frequently Asked Questions,* available at http://www.fda.gov/nanotechnology/faqs.html
97. *Ibid.*
98. National Nanotechnology Initiative, Factsheet: *What Is Nanotechnology?* Available at http://www.nano.gov/html/facts/whatIsNano.html
99. FDA, Regulation of Nanotechnology Products, available at http://www.fda.gov/nanotechnology/regulation.html
100. The Allianz Group and the Organisation for Economic Co-operation and Development (OECD), *Small Sizes that Matter: Opportunities and risks of Nanotechnologies* (June 3, 2005) at § 6.4, at 30.

101. European Commission's Scientific Committee on Emerging and Newly Identified Health Risks (SCENIHR), *Opinion on the Appropriateness of Existing Methodologies to Assess the Potential Risks Associated with Engineered and Adventitious Products of Nanotechnologies* (adopted September 28–29, 2005), at 6 (emphasis added); *ibid.* at 34.

102. See, e.g., The Royal Society and the Royal Academy of Engineering, *Nanoscience and nanotechnologies: Opportunities and uncertainties,* London (2004), at 49 (emphasis added).

103. Tran *et al.*, *A Scoping Study to Identify Hazard Data Needs For Addressing The Risks Presented By Nanoparticles and Nanotubes,* Institute of Occupational Medicine Research Report (December 2005), at 34 (emphasis added).

104. European Commission's Scientific Committee on Emerging and Newly Identified Health Risks (SCENIHR), *Opinion on the Appropriateness of Existing Methodologies to Assess the Potential Risks Associated with Engineered and Adventitious Products of Nanotechnologies,* at 6 (adopted September 28–29, 2005), at 32; Nuala Moran, *Nanomedicine Lacks Recognition in Europe,* 24 Nature Biotechnology, No. 2 (February 2006).

105. Andrew Maynard, Nanotechnology: *The Next Big Thing, or Much Ado about Nothing? Annals of Occupational Hygiene* (September 2006), at 7.

106. Maynard *et al.*, "Safe Handling of Nanotechnology," *Nature* (November 16, 2006).

107. Andre Nel *et al.*, "Toxic Potential of Materials at the Nanolevel," *Science,* 311, 622 (2006); Oberdorster *et al.*, "Principles for Characterizing the Potential Human Health Effects from Exposure to Nanomaterials: Elements of a Screening Strategy," *Particle and Fibre Toxicology* 8 (2), (2005), at 1.0.

108. *Ibid.*

109. The Royal Society and the Royal Academy of Engineering, *Nanoscience and Nanotechnologies: Opportunities and Uncertainties,* London (2004).

110. FDA, FDA Forms Nanotechnology Task Force (August 9, 2006), available at http://www.fda.gov/bbs/topics/NEWS/2006/NEW01426.html

111. FDA, FDA Nanotechnology Public Meeting, available at http://www.fda.gov/nanotechnology/meeting1010.html

112. 71 Fed. Reg. 19523 (April 14, 2006).

113. The Project on Emerging Nanotechnologies at the Woodrow Wilson International Center for Scholars, *The Nanotechnology Consumer Products Inventory Analysis.*

114. Rick Weiss, "Nanotech Product Recalled in Germany," *Washington Post,* A2 (April 4, 2006).

115. The petitioning organizations are ICTA, FOE, Greenpeace International, The Action Group on Erosion, Technology, and Concentration (ETC Group), Clean Production Action, The Center for Environmental Health (CEH), Our Bodies Ourselves, and The Silicon Valley Toxics Coalition (SVTC).

116. Available at http://www.icta.org/doc/Nano%20FDA%20petition%20final.pdf

117. Keay Davidson, "FDA Urged to Limit Nanoparticle Use in Cosmetics and Sunscreens," *San Francisco Chronicle,* May 17, 2006.

118. See 21 C.F.R. § 10.85(a).

119. See ICTA FDA Nano Petition.

120. *Ibid.*; see 58 Fed. Reg. 28195; 21 U.S.C. § 321(g)(1).
121. US Food and Drug Administration, HHS, Sunscreen Drug Products For Over-The-Counter Human Use; Final Monograph, 64 Fed. Reg. 27666-27693, 27671 (1999). FDA used the term "micronized" and the agency has not clarified whether or not this term was meant to be inclusive of nanoparticles. See also Kulinowski and Colvin, "Environmental Implications of Engineered Nanomaterials," 1 *Nanotechnology Law & Business*, 52, 53 (2004).
122. See 21 U.S.C. §§ 321(p), 355(a).
123. See, e.g., European Commission's Scientific Committee on Cosmetic and Non-Food Products (SCCNFP), *Statement on Zinc Oxide In Sunscreens* (adopted September 20, 2005), available at http://europa.eu.int/comm/health/ph_risk/committees/04_sccp/docs/sccp_o_00m.pdf (finding insufficient evidence presented for a finding of safety); see also The Royal Society and the Royal Academy of Engineering, *Nanoscience and Nanotechnologies: Opportunities and Uncertainties,* London (2004), at 73.
124. See, e.g., Hidaka et al., *In Vitro Photochemical Damage to DNA, RNA and Their Bases by an Inorganic Sunscreen Agent on Exposure to UVA and UVB radiation, Journal of Photochemistry and Photobiology*, 111, 205–213 (1997); Dunford *et al.*, "Chemical Oxidation and DNA Damage by Inorganic Sunscreen Ingredients," *FEBS Letters*, 418(1–2), 87–90 (1997); Donaldson *et al.*, *Free Radical Activity Associated with the Surface of Particles: A Unifying Factor in Determining Biological Activity? Toxicology Letters*, 88, 293–298 (1996).
125. 21 C.F.R. §§ 2.5(a) (imminent hazard), 7.45(a) (recall).
126. Michael Taylor, *Regulating the Products of Nanotechnology: Does FDA Have the Tools It Needs?* Woodrow Wilson International Center for Scholars, Project on Emerging Nanotechnologies (October 5, 2006).
127. *Ibid.*
128. *Ibid.*
129. The broader societal and bioethical implications of nanotechnology are beyond the scope of this chapter. For more information, see generally nanotechnology publications of the ETC Group, available at http://www.etcgroup.org/en/
130. Cosmetics Toiletries and Fragrances Association (CTFA), Comments on the ICTA Legal Petition.
131. See, e.g., Rick Weiss, *Nanotechnology Risks Unknown*; "Insufficient Attention Paid to Potential Dangers," Report Says, *Washington Post* A12 (September 26, 2006).
132. Andrew Maynard, *Nanotechnology: A Research Strategy for Addressing Risk*, Woodrow Wilson International Center for Scholars, Project on Emerging Nanotechnologies, (July 2006), at 23.
133. CTFA Comments on ICTA Petition.
134. Andrew Maynard, *Nanotechnology: The Next Big Thing, or Much Ado about Nothing? Annals of Occupational Hygiene*, (September 2006), at 7, Oberdorster *et al.*, "*Principles for Characterizing the Potential Human Health Effects from Exposure to Nanomaterials: Elements of a Screening Strategy,*" 2 *Particle and Fibre Toxicology*, 8 (2005), at 1.0.

135. Simply stated, the precautionary principle stands for the idea that inaction is preferable to action in circumstances where taking action could result in serious or irreversible harm. See generally Ronnie Harding and Elizabeth Fisher, eds., *Perspectives on the Precautionary Principle*, 2–3 (1999).
136. Cliff Betton, Presentation, REACH, Product Safety Assessment Ltd, *Health and Beauty America Regulatory Summit*, September 14, 2006, New York City, NY.
137. *Ibid.*
138. Piller, "Science's Tiny, Big Unknown," *Los Angeles Times* (June 1, 2006), A1.
139. Peter Hart Research Associates, *Report Findings*, Report of the Woodrow Wilson International Center for Scholars, Project on Emerging Nanotechnologies (September 19, 2006).
140. R.L. Rundle, "This War Against Germs Has a Silver Lining," *Wall Street Journal* (June 6, 2006), at D1.
141. *Ibid.*
142. Letter from Ken Kirk, Executive Director, National Association of Clean Water Agencies, to Stephen Johnson, Administrator, Environmental Protection Agency (February 14, 2006) (on file with author); Letter from Chuck Weir, Chair, Tri-TAC, to James Jones, Director, Office of Pesticide Programs, Environmental Protection Agency (January 27, 2006) (on file with author); Pat Phibbs, "Pesticides: Examining Use of Nanoscale Silver in Washing Machines as Possible Pesticide," *Daily Environment Report* (May 15, 2006), at A-5–A-6.
143. 7 U.S.C. § 136a(a).
144. Compare, FresherLonger™ Miracle Food Storage Containers, http://www.sharperimage.com/us/en/catalog/productdetails/sku_ZN020 with FresherLonger™ Miracle Food Storage Containers, HYPERLINK http://web.archive.org/web/20060208021530/http://www.sharperimage.com/us/en/catalog/productdetails/sku_ZN020. See also Rick Weiss, "EPA to Regulate Nanoproducts sold as Germ-killing", *Washington Post*, at A1, November 23, 2006.
145. A lifecycle assessment is the "systematic analysis of the resources usages (e.g., energy, water, raw materials) and the emissions over the complete supply chain from the cradle of primary resources to the grave of recycling or disposal." The Royal Society and the Royal Academy of Engineering, *Nanoscience and Nanotechnologies: Opportunities and Uncertainties*, London (2004), at 32.
146. Glenn Reynolds, *Nanotechnology and Regulatory Policy: Three Futures*, 17 Harv. J. L. & Tech. 179, 185 (2003).
147. See, e.g., Davis, J., *Managing the Effects of Nanotechnology*, Woodrow Wilson International Center for Scholars, Project on Emerging Nanotechnologies, Washington, DC (2006).

10

Navigating the Turbulent Waters of Global Colorant Regulations for Packaging

Wylie H. Royce

Royce Associates, East Rutherford, NJ, USA

10.1 Introduction

10.1.1 Authors's Note

The regulations noted in this chapter were in effect at the time of its writing. Regulations are being added and modified regularly, so please use this information as a guideline only for ensuring that your product is in compliance with the applicable laws and regulations where your product will be used. I urge you to communicate your compliance needs with your suppliers to ensure that your product meets all applicable laws and regulations.

This chapter outlines the various packaging regulations relating to colors and additives that must be adhered to when you are marketing your product in North America, the EU, and Asia.

First and foremost it must be fully understood what you are packaging. This question is more complex that it would first seem, especially if you are marketing your product internationally.

C. I. Betton (ed.), Global Regulatory Issues for the Cosmetics Industry Vol. 1, 155–164
© 2007 William Andrew Inc.

Cosmetic and personal care items typically have a number of categories that local regulatory agencies classify them into:

1. Personal care
2. Consumer goods
3. Generally recognized as safe over the counter drug

In addition, different countries and regions may define these products differently, and it is necessary to ensure your package meets regulatory compliance in every region it is marketed in.

10.2 Definitions

10.2.1 United States

If the product has no claim of efficacy then it is considered a cosmetic or personal care item.

Examples of these items are: lotion, cologne, toothpaste, and shampoo.

If your product has a claim of efficacy on the body then it is considered a drug.

Examples of drugs are: anti-dandruff shampoo, fluoride toothpaste, and anti-perspirant deodorant.

10.2.2 Canada

The definitions in Canada are virtually the same as in the United States.

10.2.3 European Union

Cosmetics, personal care and food items are all considered as "consumer goods".

10.2.4 Asia

At this time, the major Asian Countries are defining food packaging laws and no definitions exist that differentiate a cosmetic or personal care item from a drug for purposes of packaging regulations.

10.3 Regulations

One caveat to keep in mind while searching for the regulation that applies to your product, making sure you are in compliance can be very daunting, because in many cases, defined regulations do not exist for many products. If you want to be assured that you are in compliance, there may not be a regulation on the books to comply with. This subject along with other concerns will be discussed later.

10.3.1 United States

10.3.1.1 Personal Care and Cosmetics

These items are regulated by the FDA and Consumer Product Safety Commission (CPSC) in concert. The CPSC is concerned with the physical characteristics of the package, i.e., child proof lids, etc. The FDA is concerned with anything the can migrate out of the package into the product and adulterate it in any way.

The FDA has a "catchall" regulation for this, it is 21 CFR 740.10. This regulation states: *Each ingredient used in a cosmetic product and each finished cosmetic product shall be adequately substantiated for safety prior to marketing. Any such ingredient or product whose safety is not adequately substantiated prior to marketing is misbranded unless it contains the following conspicuous statement on the principal display panel: "Warning—The safety of this product has not been determined"*.

This regulation is a labeling regulation, it does not prohibit the use of any products in the container; however, from a common sense standpoint it makes sense to be in compliance.

10.3.1.2 Drugs

Any product that is defined by the FDA as a "Drug" (this includes GRASE OTC and Prescription as well) must be in compliance with 21 CFR 211.94, this applies to both the container and closure and is more restrictive than just a labeling regulation:

This regulation states: *(a) Drug product containers and closures shall not be reactive, additive, or absorptive so as to alter the safety, identity, strength, quality, or purity of the drug beyond the official or established requirements. (b) Container closure systems shall provide adequate protection against foreseeable external factors in storage and use that can cause deterioration or contamination of the drug product. (c) Drug product containers and closures shall be clean and, where indicated by the nature of the drug, sterilized and processed to remove pyrogenic properties to assure that they are suitable for their intended use.*

This can be interpreted as meaning that the package must not have anything migrate out of it into the product that can in any way alter or affect the safety or efficacy of the drug.

10.3.1.3 Other Packaging Concerns

There are other rules that packagers must be aware of in the US. They are:

California Proposition 65: This law even though it is on the books in California, in actual practice affects all packagers in the US. California has published a list of chemicals suspected to be carcinogens and updates it regularly. If any of these chemicals are intentionally added to the product, then a warning label must be placed on the product. It does not prohibit the use of the chemical, it merely has to be labeled as such.

CONEG: This law prohibits the intentional introduction of any "heavy metal" pigments or products into any plastic product that may end up in a landfill and potentially contaminate groundwater.

Note: In addition to these laws, it is of note that Minnesota is now beginning to embrace "full life cycle" packaging design similar to the EU EN13427:2004. This not only addresses the safety of the current polymer and additives for the first intended use, but also potential uses after recycling and then final environmental impact when it reaches the landfill. It also addresses package design with the goal of reducing packaging material use by 40–50%.

10.3.2 Canada

10.3.2.1 Cosmetics and Personal Care

If the product is considered either a personal care item or a cosmetic (essentially the same definition as in the US), then the package must not contain anything hazardous that may migrate out of the package into the finished product.

10.3.2.2 Drugs

If the product is a drug, it falls under the Canadian Food and Drug Act. Then it is recommended that the food contact directives be followed. As such, there is no "positives" list of additives that can be used, so if you want to have an "approval," then individual requests must be submitted to Health Canada for a letter of no objection.

10.3.3 European Union

As mentioned, the EU does not differentiate between requirements for food contact, drug, personal care, or cosmetic packaging. In general, all packaging must meet the food contact regulations, Plastics Directive. EN 71 part 3 (toy regulations) and ISO 8124-3:1997 are followed, these regulations address the potential migration of antimony, arsenic, barium, cadmium, chromium, mercury, selenium, and aromatic amines.

There exists a positives list of polymers and additives only, (colorants have been excluded) in the Plastic Directive, which is very easily complied with.

10.3.3.1 Mutual Recognition

EU countries operate under the concept of "mutual recognition". This regulation states that if a product is made within the EU to that country's specifications, then it is acceptable for use in any other EU country. For products imported into the EU, it is the country of first import that is important and if the product is accepted there, it may then be sold in any other EU country.

10.3.3.2 Colorants Lists

Both Germany and France maintain positives lists of colorants (France has a recently proposed updated list which will probably be approved in the near future) that are permitted for use in regulated packaging. Following either one of these lists will allow you to use that package in the EU. The French list appears to be the most developed and liberal use list as of this writing.

10.3.3.3 Ensuring Safety

As a general rule of thumb, if your product conforms to the Plastics Directive and uses the colorants noted on either the French or the German positives lists, or exhibits migration below the acceptable risk level of the noted chemicals, it is safe for cosmetic and drug packaging in the EU.

10.3.3.4 Other Concerns

Just as in the US there are other concerns that must be addressed by packagers to ensure that their product is not in violation of any EU laws outside of the most well-known food contact regulations they are most notably as follows.

The Seventh Amendment: This law prohibits the introduction of substances known to cause cancer, birth defects, and certain allergens into a product. Most notable is dibutylphthalate which may be used as a plasticizer in PVC. The key to ensuring compliance with the Seventh Amendment is to make sure there is no migration of additives or potentially hazardous chemicals to the finished product. This can be challenging in fragrances. Fragrances typically have several oils, solvents, etc. in their formulations. All of these products dramatically increase the potential of migration out of the package and must be checked very carefully in accordance with the package's final intended use.

Full Life Cycle Directive: The EU is not only very concerned with the initial intended use of packaging but is also now embracing the theory that all packaging should have several uses before in ends up in the landfill and should be minimized in size in order to minimize environmental impact. This Directive is just advisory in nature, but can be a harbinger of things to come for packaging manufacturers in the years to come.

The Directive is: EU Directive 92/62/EC Packaging and Packaging Waste Directives. EU Full Life Cycle Product Law. Updated under the "Umbrella Standard" EN 13427:2004.

10.4 Choosing the Right Colorants for Your Product

The resin that is used in your package will determine the colorant formulary for your product. Dyes and pigments cannot be used interchangeably between different resins.

10.4.1 Dyes

Dyes are bright clean transparent colorants. Many effects that are achieved with dyes cannot be achieved with pigments. But because dyes are soluble in oil and solvents, they are limited to being only used in non-olefin resins. In addition, dyes also have the ability to occasionally migrate from non-olefin resins in certain instances, so migration testing may be needed to meet regulations in the EU, and drug packaging regulations in the US. Migration testing, however, is quite inexpensive and can be easily performed by any competent laboratory.

10.4.2 Pigments

Pigments remain in an insoluble state, so they are not likely to migrate from any resin and thus are more universally useable than dyes. But pigments with the exception of a few unique products cannot produce clear, clean transparent effects and are typically much duller in appearance than dyes. If your package is an olefin-based product, or contains and olefin-based modifier, you *must* use pigments and accept the affect that they are capable of producing.

10.4.3 FD&C Colorants

There is a common misconception that FD&C colorants must be used in cosmetic packaging. This is simply not the case. In fact, FD&C colorants are typically very expensive, limited in scope and weak in tint strength. In short, do not specify their use for packaging applications.

10.4.4 Migration

When you design your package and color effect, the mantra to keep in mind is that if nothing is shown to migrate above an acceptable risk level, it will be acceptable for use throughout the world. If migration of a chemical exists, then the FDA, Health Canada, and EU will be concerned with what is migrating from the package and what affect it may have on the product it is packaging.

10.5 Meeting the Design Challenge

When you are designing your package, it is suggested that to ensure legality you consider the following steps in your design process:

1. Determine what the product will be classified as, in the market you will be selling it. Will it be a cosmetic, personal care item, or drug?

2. What functional properties does the package need to have, so will you be limited to only pigments versus any combination of dyes and pigments? The resin will dictate the available effects and colors.
3. What are the end markets going to be, US, Canada, EU or Asia, or any combination of them?
4. Effect versus cost, what is the driving parameter for your product? Brighter and hotter colors will cost substantially more, but the cost differential can be minute when taken in the context of the entire product cost.

10.5.1 Risk Assessment

Virtually every regulation is based on the theory of risk assessment. In short, what are the associated risks of something hazardous migrating from the package into the finished product? If migration below an acceptable level can be demonstrated, then no real risk exists and the various agencies will not be concerned with your package.

10.5.2 The Quandry of Non-Regulation Regulations

How do you ensure legality or acceptability of a package when no definite regulations exist? Unfortunately, no simple answer exists, because something that may be acceptable for food contact packaging (the most stringent in the US) may potentially have a skin contact or allergic reaction causing a bio-compatibility problem.

Currently, the only advice that can be proffered is, if you want to ensure that your package is safe, make sure that nothing will migrate from it. Resins have undergone rigorous testing to assure bio-compatibility and non-allergic reactions. So in theory, if nothing is going to come out of the resin, there will not be a reaction. This is a practical approach, but keep in mind, it is impossible to certify a product to a regulatory standard that does not exist in the first place!

10.5.3 Allergens

Most systemic allergic reactions are food related, not an FDA concern with cosmetics. If your product is categorized as a drug, you should ensure that there are no wheat, milk, shellfish, peanuts, tree nuts, eggs, soybeans, or fish used in any of the components of the package that could reasonably migrate to the product. Skin allergy is a major issue with cosmetics, the major causes being preservatives and fragrance ingredients. The likelihood of a component migrating from packaging to give a concentration in the product sufficient to generate a dermal allergic response is so low as to be insignificant.

10.5.4 Testing

It is not recommended that extensive testing be insisted on for these types of situations. First, the laboratory will not know what standard to test to since none exists. Second, it can become extremely expensive to try to test for every possible scenario.

10.6 Importing Packaging

It is extremely important to know and manage your risks. Packaging that is contracted for is typically manufactured to meet the regulations that exist in the country of manufacture. In short, the EU, Canadian, and US laws are more stringent than Asian laws. Most Asian countries are just now in talks to develop food contact packaging regulations that are as stringent as other areas. Most Asian regulations simply state that the product must be non-hazardous in nature, and non-hazardous in nature currently has an acceptable "heavy metal" content level in food contact packaging.

So while the package may meet local regulations, it can possibly be way out of spec when imported into your country.

It is imperative that if you are importing packaging and want to ensure regulatory compliance that your supplier certify that all of the components of the package meet the laws of your country.

10.6.1 Liability Issues

There is a little known liability exposure that exists in packaging certification that should be noted,

That exposure is, *any certification that is written is viewed by the courts as a separate contract to any others and exposes the supplier to unlimited liability!* This has been tested and upheld in the courts.

To avoid this unlimited exposure, you just need to either state on the certification that this certification is subject to your normal terms and conditions of sale, or explicitly state the liability limitations on the sheet.

10.7 Regulations at a Glance

Locale	Cosmetic	Drug
US	Non-hazardous CSPC and FDA Regulated See FDA 21 CFR 740.10	Any drug as defined by FDA FDA CFR 211.94
Canada	All ingredients non-hazardous Regulated by Consumer Labeling and Packaging Act	Drug definition similar to US Regulated by Food and Drug Act. No positives list, color must be submitted to Health Canada for letter of no objection.
EU	Non-hazardous comply with: Plastics Directive, Food Contact regulations, ISO 8124-3: 1997, EN 71 Part 3 Toy Regulation, EN 13427:2004 Umbrella	Not applicable, as cosmetics and personal care items are not classified as drugs
Asia	Non-hazardous No other regulations defined	Not applicable to cosmetics and personal care items

Index

Absorption, 104–106
in vitro oral absorption studies, 104
ACD (allergic contact dermatitis), 98
Acute toxicity, 88–92
ADME (absorption, distribution,
metabolism, and excretion), 90, 103
Allergens, 162–163
Allylisothionate, 7
Alpha-isomethyl lonone, 68
Amyl cinnamal, 68
Animal carcinogens, 7
allylisothionate, 7
benzyl acetate, 7
estragole, 7
glutamyl-*p*-hydrazinobenzoate, 7
p-hydrazinobenzoate, 7
limonene, 7
5- and 8-methoxypsoralen, 7
safrole, 7
sesamol, 7
Animal testing, 84
Anise alcohol, 68
Asia, 156
Atropine, 10
Australia, regulatory developments in,
28–29
composition of the product, 28
cosmetics category, 28
proposed use of the product, 28
therapeutic products category, 28

Benzyl acetate, 7
Benzyl alcohol, 68
Benzyl benzoate, 68
Benzyl cinnamate, 68
Benzyl salicylate, 68
Bioaccumulation, of nanoparticles, 131
BSE (bovine spongiform
encephalopathy), 30–31
Buckyballs, 118
Business engagement, in global regulatory
strategy, 59
Butylatedhydroxyanisol (BHA), 63

California
California Proposition 65, 67
California Safe Cosmetics Act
of 2005, 67
regulatory developments in, 32
Canada, 156, 158–159
cosmetics and personal care products,
regulation, 158
drugs, 159
Canada, regulatory developments in,
21–33
bilingual language requirement, 22
Category IV drugs, 24
cosmetics, 24
Health Canada agency, 23
mandatory labeling requirements, 23
Natural Health Products (NHP)
program, 24
Cancer/Carcinogenicity, 100–101,
See also Carcinogens
causes, 1
deaths, 3
definition, 100
regulation, 2
as risk, 4
Carbon fullerenes (C_{60}), 118
Carboxyalhehyde, 68
Carcinogens, *See also* Animal
carcinogens
chemicals, 4
classification, 6–7
coffee, 6
human carcinogens, 8
naturally occurring food constituents
causing, 7
tomatoes, 6
CASE (Computer Automated Structure
Evaluation), 98
Chemicals, risks of, 4
China, regulatory developments in, 29–31
registration process, 30
Cinnamal, 68
Cinnamyl alcohol, 68

Citral, 68
Citronellol, 68
CMA (Chemical Manufacturers
  Association), 70
Colorant regulations for packaging,
  155–164
  design challenge of, 161–163, *See also*
    Design challenge
  Directive, 160
  Full Life Cycle Directive, 160
  importing packaging, 163–164
  products, definitions, 156
  regulations, 157–160
    Canada, 158–159
    EU, 159–160, *See also* EU
    US, 157–158, *See also* US
  Seventh Amendment, 160
Colorants, choice of, 160–161
  dyes, 160–161
  FD & C colorants, 161
  migration, 161
  pigments, 161
Communication plan, in global regulatory
  strategy, 59–60
Consumer products
  nanomaterials in, 123–125
    maturation, measures, 123–125
Consumer products, restricted substances
  in, 71–81
  3 Cs to compliance, 73
  best practices, 77–79
  emerging globally restricted substances
    regulations, 73–74
  global compliance directives, steps to
    meeting, 75–77
    communication, 76
    compliance management system,
      setting, 75
    documentation and proof collection, 76
    measuring risk, 76
    strategy, preparing, 75
  strategy definition, 74–77
Coumarin, 68
CPSC (Consumer Product Safety
  Commission), 66
Cross-functional processes, systems,
  and tools
  in global regulatory strategy, 59

DDT, 38
DEREK (Deductive Estimation of Risk
  from Existing Knowledge), 97–98
Design challenge of package, 161–163
  allergens, 162–163
  quandry of non-regulation
    regulations, 162
  risk assessment, 162
  testing, 163
Distribution
  in vitro modeling of, 106–107
D-limonene, 68
Draize tests, 10
Dyes, 160–161

EINECS (European Inventory of Existing
  Chemical Substances), 37
ELINCS (European List of New Chemical
  Substances), 37
Environmental risks of nanomaterials
  in personal care products, 130–132
    bioaccumulation, 131
    in soil, 130
EpiDerm model, 94
EPISKIN model, 94
Estragole, 7
EU (European Union), 156, 159–160
  colorants lists, 159
  cosmetic legislation and animal testing,
    68, 84–85
  ensuring safety, 159–160
  fragrance ingredients requiring labeling
    under, 68
  mutual recognition, 159
  regulation of cosmetics, 11–12
Eugenol, 68
Evernia prunastri extract, 68
Ex vivo testing, 96
Excretion, 108
Eye
  irritation, 95–96
  risk assessment on, 17

Farnesol Butyl methylpropianal
FD & C colorants, 161
FDA (Food and Drug
  Administration), 119
  FDA's regulatory stance

on nanotechnology and nanomaterial personal care products, 132–133
nanomaterial oversight developments from, 135–138
US, 65
FHSA (Federal Hazardous Substances Act), 66–69
FIFRA (Federal Insecticide, Fungicide, and Rodenticide Act), 143
FoE (Friends of the Earth), 125
Formaldehyde, 63–64
Fragrance ingredients requiring labeling under EU, 68

Geraniol, 68
GJIC (gap junction intercellular communication) assays, 101
Global chemical compliance, 71–81
Global regulatory strategy, developing, 55–62
  business engagement, 59
  communication plan, 59–60
  cross-functional processes, systems, and tools, 59
  global launch, 60–62
  local knowledge, 58
  organizational design, 57–58
  regulatory compliance for competitive advantage, 56–60
  targeted capability, 58
Globalization
  and REACH regulation, 50
Glutamyl-*p*-hydrazinobenzoate, 7
Guinea Pig test methods, 97

Hair treatment products, 64
Health risks of nanomaterials
  in personal care products, 125–130
    nanotoxicity, 126–127
    public at large, 128
    skin penetration, 128–129
    unprecedented mobility, 127–128
    worker and workplace risks, 129–130
Hexyl cinnamal, 68
Hydroquinone, 69
p-Hydrazinobenzoate, 7
Hydroxycitonellal, 68
Hydroxyisohexyl-3-cyclohexene, 68

ICTA (International Center for Technology Assessment), 135–136, 138
Importing packaging, 163–164
In silico systems, 93, 107
In vitro toxicology for cosmetics
  acute toxicity, 88–92
  ADME outcomes, 90
  carcinogenicity, 100–101
  EpiDerm model, 94
  EPISKIN model, 94
  EU cosmetic legislation and animal testing, 84–85
  eye irritation, 95–96
  Guinea Pig test methods, 97
  histological evaluation, 96
  in vitro genotoxity tests, 100
  in vitro replacement tests, 107
  in vivo and, comparison, 95, 110
  LLNA (Local Lymph Node Assay), 97
  NRU (neutral red uptake) measurement, 90–91
  regulatory requirements, biological limitations, 83–114
  reproductive/developmental toxicity, 101–103
  skin corrosion, 92–93
  skin irritation, 93–95
  skin sensitization, 96–99
  toxicokinetics, 103–108, *See also separate entry*
India, regulatory developments in, 31
Ingestion, risk assessment on, 17
Inhalation, risk assessment on, 17
International Agency for Research on Cancer (IARC), 6
Intertek, 73–74, 76, 81
Iso-eugenol, 68
Isolated organ tests, 95

Japan, regulatory developments in, 25–28
  approval process, 25–26
  cultural issues, 27
  labeling, 26
  Ministry of Health, Labor and Welfare (MHLW), 25
  quasi-drugs, 26–27

Labeling, of cosmetics, 97
  in Canada, 26
  in Japan, 26
Langerhans cells
  in testing, 98–99
Lanolin alcohol, 64
LD50s (lethal dose), 88, 91
Lifecycle, 143–144
Limonene, 7
Linalool, 68
LLNA (Local Lymph Node Assay), 97–99
Local knowledge, global regulatory
  strategy and, 58

Mad Cow Disease, 30–31
Metabolism to *in vitro* testing, 107–108
5- and 8-Methoxypsoralen, 7
Methyl and propyl paraben, 64
Methyl-2-octynoate, 68
MSDS (Material Safety Data Sheets), 69

Nanotechnology/Nanomaterials
  'nano' meaning, 121
  buckyballs, 118
  definition and description, 119–123
    engineered/manufactured
      nanoparticle, 120
    nanomaterial, 120
    nanoproduct, 120
    nanoscale, 120
    nanoscience, 120
    nanotechnologies, 120
  environmental risks of, 130–132,
    See also Environmental risks
  FDA's regulatory stance on, 132–133
  huge EHS unknowns, 131–132
  human health risks of, 125–130,
    See also Health risks
  in consumer products, 123–125,
    See also Consumer products
  manufactured and engineered
    nanomaterials vs. natural
    nanoparticles, 121–122
  maturation, measures, 123–125
  nano-specific testing paradigms,
    133–135
  oversight developments from FDA,
    135–138

personal care products, 117–153
predicted developmental stages, 122–123
recommendations for government and
  industry, 138–145
  EHS research, supporting, 139
  lifecycle, 143–144
  meeting EU standards, 140–141
  past, learning from, 144–145
  transparency, 141–142
  unknowns and fundamental
    differences of nanomaterials,
    acknowledging, 139–140
NEPA (National Environmental Policy
  Act), 137
NHP (Natural Health Products)
  program, Canada, 24
NRU (neutral red uptake)
  mesurement, 90–91

OSHA (Occupational Safety and Health
  Administration), 66–69

Packaging, colorant regulations for,
  155–164, *See also* Colorant regulations
Personal care products, 125–133, *See also*
  under Nanotechnology/Nanomaterials
Pigments, 161
Polyethylene glycol, 64
PPB (plasma protein binding), 105–106
Precautionary principle, 9, 52, 87,
  140–141
Propylene glycol, 64
Purchaser of cosmetics
  expectations, 10
PVC, 38

QSAR (Quantitative Structure-Activity
  Relationship), 97–98
Quandry of non-regulation
  regulations, 162
Quasi-drugs, 26–27
Quaternium 15, 64

REACH (Registration, Evaluation,
  Authorisation of Chemicals) regulation,
  49–54, 140–141
  aims, 39–44
  description, 51–52

effect on cosmetics industry, 52
of EU, 35–47
exemption for cosmetics, 45
exemptions, 44–46
implications, 72
new rules, 50
overview, 38–39
for polymers, 45
pre-registration, 41–43
reason for, 36–38
registration activities required by,
  timeline for, 42
registration, 46–47
SIEF, 43–44
working of, 52–53
Regulations/Regulatory developments,
  See also individual country names
Australia, 28–29
California, 32
Canada, 21–33
China, 29–31
for colorants packaging, 157–160
India, 31
Japan, 25–28
in risk assessment, 14
trade alliances, 31–32
Reproductive/developmental toxicity,
  101–103
Restricted substances in consumer
  products, 71–81, *See also*
  Consumer products
Risk (R) phrases, 6
Risk assessment and cosmetics, 1–20
  cancer, causes of, 1
  chemicals, 4
  contents, 17–18
  guidance, 15
  regulations, 14
  risk perception and regulation, 1–10
  risk, hazard, and exposure,
    relationship, 12–13
  safety assessment, 16–17, *See also*
    separate entry
  safety assessor, role of, 14–15
Risks, of nanomaterials
  environmental risks, 130–132
  human health risks, 125–130
  workplace risks, 129–130

Safety assessment, 16–17
  physical nature of the product, 16
  products usage, 16
  safety assessor, role of, 14–15
  traditional methods, 10–11
Safrole, 7
SCCNFP (Scientific Committee for
  Cosmetics and Non-Food Products), 15
SCCP (Scientific Committee on
  Consumer Products), 14
Sesamol, 7
SIEF (Substance Information Exchange
  Fora), 43–44
Skin
  corrosion, 92–93
  irritation, 93–95
  penetration risk, of nanomaterials,
    128–129
  risk assessment on, 17
  sensitization, 96–99

TGA (Therapeutic Goods
  Administration), 125
Thalidomide, 102
Tissue partitioning, 105
TOPKAT (Toxicity Prediction by
  Komputer Assisted Technology), 97–98
Toxic chemicals in cosmetics, 64
  ointment bases, 64
  preservatives, 64
Toxicity and regulatory requirements in
  the US, 63–70, *See also under* US
Toxicokinetics, 103–108
  absorption, 104–105
  distribution, 105–106
  excretion, 108
  in silico computer modeling, 107
  metabolism, 106–107
  PPB (plasma protein binding),
    105–106
  tissue partitioning, 105
Transparency, 141–142
Tributyl tin, 38

US (United States), 156–158
  drugs, 157–158
  personal care and cosmetics,
    regulations, 157

US (United States) (*contd.*)
  toxicity and regulatory requirements of
    cosmetics in, 63–70
    butylatedhydroxyanisol (BHA), 63
    California proposition 65, 67
    California Safe Cosmetics Act of
      2005, 67
    cosmetics regulatory requirements,
      future, 68–70
    EU Cosmetics Regulation, 68

    formaldehyde, 63
    hair treatment products, 64
    ingredient declaration, 67
    regulatory requirements for
      cosmetics, 65–67
    toxic chemicals in cosmetics, 64
    warning statement, 66

Worker and workplace risks, of
  nanomaterials, 129–130

Breakthroughs in Personal Care and Cosmetic Technology Series

Series Editor: Meyer R. Rosen
Interactive Consulting, Inc., East Norwich, New York

Delivery System Handbook for Personal Care and Cosmetic Products: Technology, Applications, and Formulations, Meyer R. Rosen, ed., 978-0-8155-1504-3, 1104 pp., 2005

Global Regulatory Issues for the Cosmetics Industry, Vol. 1, C. I. Betton, ed., 978-0-8155-1567-8, 192 pp, 2007

Forthcoming 2008
Skin Aging Handbook
Edited by Nava Dayan

Cosmetic Applications of Laser and Light-Based Systems
Edited by Gurpreet Ahluwalia

Forthcoming 2009
Handbook of Non-Invasive Drug Delivery Systems
Edited by Vitthal S. Kulkarni

Nutritional Cosmetics
Edited by Aaron Tabor and Robert M. Blair

Of Related Interest

Cosmetic and Toiletry Formulations Database (**CD-ROM**), Ernest W. Flick, 978-0-8155-1507-4, 2005

Pharmaceutical Manufacturing Encyclopedia, 3rd Edition, in 4 volumes, 978-0-8155-1526-5, 3856 pp, 2006

Printed and bound by CPI Group (UK) Ltd, Croydon, CR0 4YY

08/05/2025

01864832-0001